A FIELD GUIDE TO

SPIDERS & SCORPIONS

OF TEXAS

A FIELD GUIDE TO

SPIDERS & SCORPIONS OF TEXAS

JOHN A. JACKMAN

A Gulf Publishing Book
TAYLOR TRADE PUBLISHING
Lanham • Dallas • New York • Toronto • Oxford

A GULF PUBLISHING BOOK
PUBLISHED BY TAYLOR TRADE PUBLISHING
An Imprint of the Rowman & Littlefield Publishing Group
4501 Forbes Boulevard, Suite 200
Lanham, MD 20706

Distributed by National Book Network

Library of Congress Cataloging-in-Publication Data

Jackman, J. A.
A field guide to spiders & scorpions of Texas / John A. Jackman.
p. cm. – (Gulf Publishing field guide series)
Originally published: Houston, Tex. : Gulf Pub., c1997, in series: Texas Monthly field guide series.
Includes bibliographical references (p.) and index.
ISBN 0-89123-048-3
1. Spiders–Texas–Identification. 2. Scorpions–Texas–Identification. 3. Ticks–Texas–Identification. 4. Arachnida–Texas–Identification. I. Title. II. Title: Field guide to spiders and scorpions. III. Title: Spiders & scorpions. IV. Series.
QL458.4.J33 1999
595.4'4'09764–dc21 99-13547
CIP

Color photographs by John A. Jackman unless otherwise noted. Contributing photographers: W. L. Sterling, W. D. Sissom, B. M. Drees, G. McIlveen, and P. J. Reynolds.

Printed in the United States of America.

To my wife and sons, whom I neglected for extended periods while preparing this book; to my parents, who always encouraged my interests; and to R. J. Sauer, who introduced me to spider taxonomy.

Contents

Acknowledgments

I wish to thank a number of people who have helped with this book. D. Allen Dean, more so than anyone, has been a local authority for this project. He provided literature, manuscript reviews, encouragement, and identifications that were essential in making this publication possible. Most importantly, he provided a copy of an unpublished catalog of Texas spiders that was most helpful in the preparation of this book.

H. R. Burke and E. G. Riley provided me free access to the collection of insects and spiders at Texas A&M University. This valuable resource was very useful in the completion of this project. I must also thank E. G. Riley for accompanying me on spider-collecting trips, pointing out some species that I might have otherwise overlooked, and offering suggestions on the format of the book.

Publications by R. G. Breene, W. L. Sterling, D. A. Dean, and M. Nyffeler were most helpful in my understanding of the spider fauna in Texas. This group of scientists in the Department of Entomology at Texas A&M University have contributed greatly to the knowledge of spiders by their work in agricultural ecosystems. Their publications are important because they provide information specifically from Texas.

P. D. Teel supplied ticks to photograph, literature on this group, and a review of that section of the book. W. D. Sissom wrote and supplied most of the information on scorpions.

Several people contributed spiders or allowed me to photograph specimens they had found. Specifically, I thank Bill Johnson, James Hannah, and Barbara Dott. I sincerely hope that I have not overlooked anyone in this regard.

D. A. Dean, R. G. Breene, W. B. Peck, D. B. Richman, and W. D. Sissom provided reviews of the manuscript which improved the content and quality. I thank R. E. Frisbie and P. J. Hamman for their interest and support throughout this project.

R. G. Weber has single-handedly elevated my photographic expertise (and that of many Extension Entomologists in Texas), thereby making the task of communication in this book feasible. While most of the photographs were taken by me, contributions by W. L. Sterling, B. M. Drees, and G. McIlveen are also included. W. D. Sissom supplied most of the photographs of scorpions. I have also included a number of photographs from P. J. Reynolds of Detroit, Michigan, that were supplied to S. G. Wellso many years ago with the hope that the slides could be used for scientific studies.

And finally, thanks go to Toni King for editing the manuscript and shepherding it through the production process.

PREFACE

Spider taxonomy has had a history of relatively few but intense authors. Publications by Comstock, Emerton, Kaston, Gertsch, Levi, Roth, and Platnick are monumental in describing the biology and nomenclature of spiders in North America. Works by these key authors make up the core of the material from which this book was written. Of course, there are many other authors who have contributed more specifics of spider biology. This list would be too long to attempt here.

The last catalog of spiders in Texas was published by Vogel (1970a), who listed 560 species names. The list is now closer to 900 species in Texas, and more are likely to be added. There have been several spider studies in Texas directed at specific locations, spider families, or agricultural ecosystems. Studies have been conducted at Nacogdoches (Brown 1974), Stephenville (Agnew et al. 1985), Dallas (Jones 1936), Austin (Vogel and Durden 1972), and Galveston Island (Rapp 1984). Spiders as members of the cave fauna have also been studied (Reddell 1965, 1970). Studies on specific families include crab spiders (Cokendolpher et al. 1979), Gnaphosidae (Zolnerowich and Horner 1985), and jumping spiders (Carpenter 1972). Agroecosystems that have been studied include citrus (Breene et al. 1993a), corn (Knutson and Gilstrap 1989), cotton (Dean et al. 1982, Dean and Sterling 1987, Breene et al. 1993b), guar (Rogers and Horner, 1977), peanuts (Agnew et al. 1985), pecans (Bumroongsook et al. 1992, Liao et al. 1984), rice (Woods and Harrell 1976), and sugar cane (Breene et al. 1993c).

Four limitations to this publication should be acknowledged. First, this book is intended to be a field guide with emphasis on representative species, large spiders, spiders with potentially hazardous venom, unusual spiders, and those most commonly encountered. Consequently, I have included some spiders that are large and conspicuous but not really very abundant. Conversely, some of the most common small or nocturnal spiders, especially Linyphiidae, are almost entirely ignored. This book should help you identify some families of spiders and a few of the more easily recognized species. Representatives of some related groups of arachnids, such as ticks, scorpions, whipscorpions, windscorpions,and pseudoscorpions are also included.

Second, this book will not make you an instant spider taxonomist. Species identification of spiders invariably requires inspection of the genitalia using a good microscope and the right literature. An experienced taxonomist who specializes in a particular family of spiders may be needed to make the identification. Spider taxonomy is always progressing, and recently, some major changes in family names (Coddington and Levi 1991) and generic placement have made it difficult to track names in the older literature. Many more changes will certainly occur. For keys to families and genera, the works by Roth (1993) and Kaston (1978) remain the most useful general publications for the United States. However, the current placement of genera and family composition has changed considerably since Kaston (1978).

Third, the biology of individual spider species is often inferred from the known biology of other members of the family. The number of spider species that have well-known biology is limited. Gertsch (1979) summarizes the biology of spiders in his book.

Fourth, the geographic distribution of individual spider species can be vague. Spiders are not the most commonly collected organisms by scientists, and the necessity to store them in alcohol limits the interest in and feasibility of maintaining a large collection. The ability of spiders to distribute themselves by "ballooning" probably makes their geographic distribution less important than the microclimate situations they need in which to thrive. Additional distribution records of spiders in Texas are expected. The spider fauna of south Texas certainly has affinities to Mexico and more tropical regions.

The family, genus, and species names in this book follow the convention of Platnick (1989, 1993) and his predecessor Brignoli (1983). There are a few names that have been changed since Platnick (1993) that are included in the checklist. The common names follow the convention of *Common Names of Arachnids* (Breene et al. 1995) published by The American Tarantula Society. Common names from other sources and scientific names that have been associated with certain species are listed under a category entitled "outdated and unofficial names" and are not meant to have scientific standing.

The descriptions and information on biology and distribution are taken primarily from Kaston (1978) and Breene et al. (1993b) and are augmented with information from other scientific literature and the aforementioned catalog by D. A. Dean.

John A. Jackman, Ph.D.
Texas A&M University
College Station, Texas

HOW TO USE THIS BOOK

This book consists of photographs of common spiders and their relatives, data and names summarized in tables and lists, and narratives about orders, families, genera, and species in the class Arachnida.

Use the photographs (**numbers in text correspond to numbers–families–in color section; letters identify genus or species**) and descriptions to identify some of the more common species. Use the tables and lists to help put the diversity of spiders in perspective. Use the narratives to learn about specific orders, families, genera, and species. When studying spiders, try to use all of the information available to you.

There are four basic ways to identify spiders: 1) ask an expert; 2) compare a specimen to authoritatively identified specimens; 3) compare the specimen to the original descriptions; and 4) use identification keys or other short-cut methods.

Spider experts and identified collections are few and not readily available. Original descriptions of spiders are tedious to use and require a good knowledge of spider anatomy, a considerable collection of literature, and experience.

Identification keys are often difficult to use and demand an understanding of specific characters–which may be hard to see–and necessitates that a preserved specimen is at hand. Specialists in spider taxonomy certainly rely on identification keys; however, they seldom find it necessary to key a specimen all the way through family, genus, and species. More often, they recognize a family or genus at a glance or are able to start at least part way through the identification keys. Thus, specialists use recognition as the first step to move through the identification process.

Short-cut methods, such as quick recognition characters, are provided in this book, although they lead to less definitive identifications than an authority would provide. However, with observation and careful deduction, one can become reasonably confident using quick recognition methods even though a higher level of taxa may be all that is achieved. A family name or genus name may actually be very difficult to place with certainty. It is always better to be cautious with identification than to jump to conclusions quickly. Nevertheless, a family name or a genus name still provides useful and often sufficient information. Problem areas of identification are discussed in the book.

Identification is as much a process of elimination as one of recognition. Like Sherlock Holmes on a case, if you eliminate the possibilities, what remains is the likely solution.

Expect a few exceptions to the information in this book. Some spider names will continue to change, as will the placement in families.

While an understanding of spider nomenclature is a challenge and a worthy goal, it should not be an end point. There is still much to be learned about the basic biology of spiders and their role in our ecological communities. Careful observation of spiders can provide useful information on spider biology that is poorly known.

This book is intended to provide the reader with a perspective of the number and types of spiders in Texas. I hope that the information will allow the reader to make rapid progress in recognizing some of the more common spiders and inspire him or her to further investigate them. Provided are some of the key reference sources for the more intense study of spiders.

Order Araneae—Spiders

Spiders—An Introduction

Spiders evoke fear in many people. With few exceptions, these fears are not justified. In reality, spiders share our homes, gardens, and parks, quietly living an independent existence. Most are hardly noticed or simply ignored by man unless they are abundant or large. Spiders are common predators in terrestrial ecosystems and serve an important role in the balance of nature. The sheer number of spiders and their behavior as general predators help maintain the world's insects at reasonable numbers.

Spiders feed primarily on living insects and other small arthropods. They are generally more particular about the size of the prey than the type of organism. However, some spiders do show food preferences. Some larger spiders may feed occasionally on small mammals, birds, fish, frogs, lizards, or even snakes. Spiders are cannibalistic and often eat their brood mates soon after they hatch. Recent studies have shown that spiders also feed on arthropod eggs, dead animals, and pollen, but these items are probably only a small part of their diet.

Spiders are found primarily in terrestrial habitats but can occupy practically every type of habitat. Some of the pisaurid spiders and lycosid spiders are common along waterways and even venture out on the surface of the water, which is surprising because they weigh much more than insects that skate on the surface film. These spiders feed on aquatic organisms on or even below the surface. Some occasionally feed on small fish.

Spiders produce silk with which they build webs to capture prey. The flat, circular orb webs are perhaps the best recognized spiderwebs. The irregular, messy cobwebs in the corners of houses and the flat funnel webs in the grass along roadsides are some of the other types of webs spiders build to capture prey. Spiders use silk as drag lines to stop themselves when they leap into space or fall from a perch. They also use silk to form retreats or temporary shelters, to hold females while mating, and to shelter their young. Male spiders build sperm webs, deposit sperm onto the web from genital pores on the front of the abdomen, and then transfer the sperm to their pedipalps before they mate. Perhaps the most spectacular use of silk is in a process called "ballooning," which refers to flying through the air on a strand of silk—a common practice of young spiderlings. This is one of the main ways in which spiders migrate to new locations to colonize.

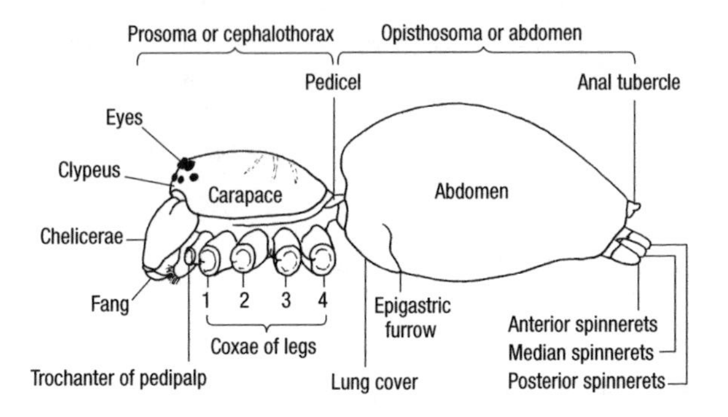

Figure 1a. Lateral view of a spider with legs and pedipalps removed. *(Redrawn from Kaston 1978.)*

Spiders are a large group of diverse species. There are over 35,000 species in the world (Coddington and Levi 1991) and over 3,000 species in the United States (Young and Edwards 1990). There are records for nearly 900 species in Texas, but the number will undoubtedly climb as we continue to study this group.

*Anatomy**

Body. The spider body consists of two parts: the cephalothorax and the abdomen (Figures 1a, 1b, and 1c). Spider size is usually reported by the body length from the front of the carapace (top of the cephalothorax) to the end of the abdomen. This measurement is more consistent than one that includes legs that are often bent, folded, or lost in preserved material. The cephalothorax serves as the head and the thorax combined. The general shape of the cephalothorax is useful in identifying spider families and, to some extent, genera. The cephalothorax shape should be viewed from both the top and the side for best identification.

At the front of the cephalothorax are the mouth parts which consist of chelicerae on the front of the head with fangs at the tip end. Next to their chelicerae, spiders have pedipalps which may appear more like a pair of small legs. Pedipalps are used to manipulate food and, in males, to mate. Continuing around the cephalothorax are the eight legs attached to the sides. The proportions of body size to leg length is sometimes given by the PT/C index. The PT/C index is the length of the patella plus tibia divided by the length of the carapace times 100. A PT/C index under 100 indicates a stout, short-legged spider, while an index of 500 indicates a spindly, long-legged spider (Roth 1993).

The abdomen is attached to the rear of the cephalothorax by a narrow connection–the pedicel–giving spiders the distinct two-body-part appearance.

---

* This section is modified from Kaston 1978.

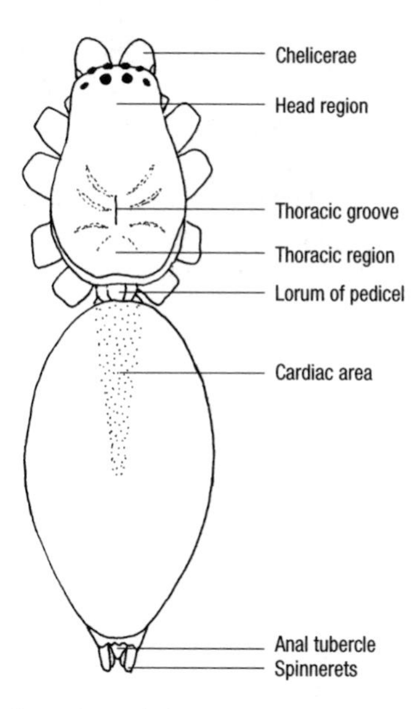

Figure 1b. Top view of a spider with legs and pedipalps removed. *(Redrawn from Kaston 1978.)*

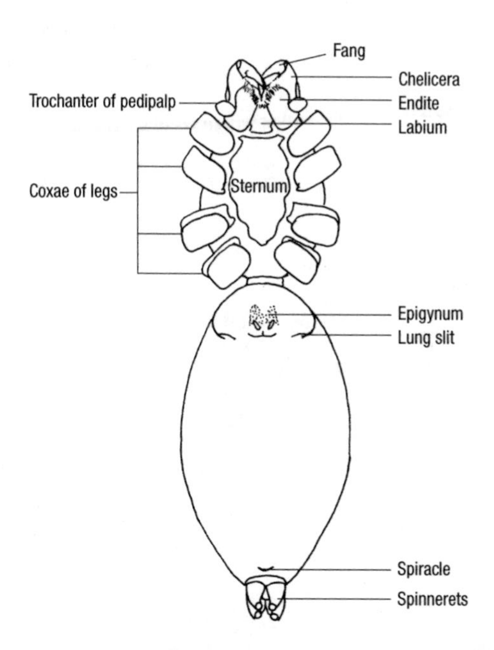

Figure 1c. Ventral view of a spider with legs and pedipalps removed. *(Redrawn from Kaston 1978.)*

Legs. Spider legs have several segments, which are named from the body outward as coxa, trochanter, femur, patella, tibia, metatarsus, and tarsus (Figure 2). The coxa and trochanter are small segments; the femur is usually the largest. At the tip end of the spider leg (the end of the tarsus) are two or three claws and various small hairs that may form a pad (Figures 3a, 3b, and 3c). The number of claws, the presence of claw tufts, and the configuration of spines and hairs on the legs are used to identify spiders.

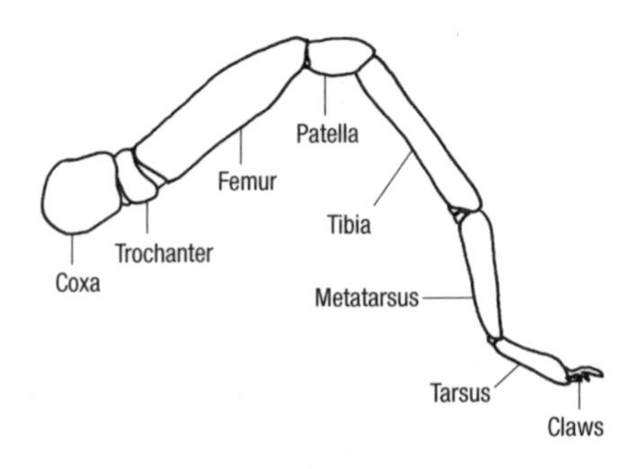

Figure 2. Leg of a spider. *(Redrawn from Kaston 1978.)*

Some spiders also have very fine hairs, or trichobothria, that project perpendicularly (or nearly so) from the legs and which are also diagnostic family characters. Some authors use a leg formula to describe the relative length of legs. This formula lists the leg number starting with the longest to the shortest. Thus, a leg formula of 1432 has the first leg longest and the second leg shortest.

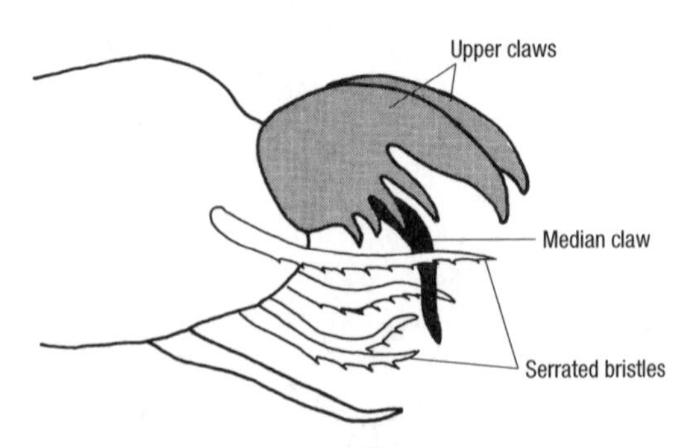

Figure 3a. Tip of hind tarsus of a three-clawed spider. Note that hairs are not shown. *(Redrawn from Kaston 1978.)*

Figure 3b. Tip of tarsus of a two-clawed spider without claw tufts. Note that hairs are not shown. *(Redrawn from Kaston 1978.)*

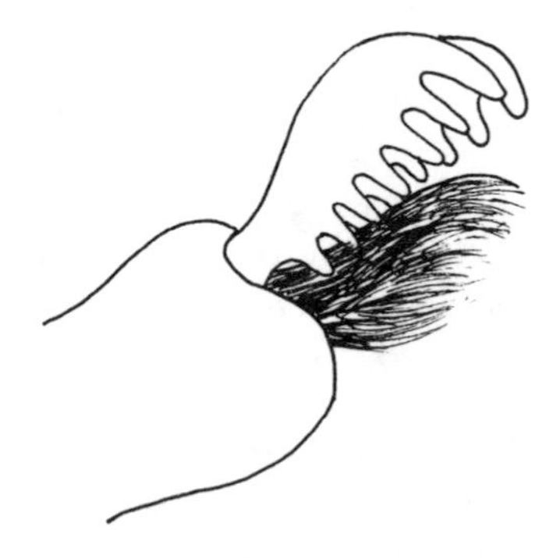

Figure 3c. Tip of tarsus of a two-clawed spider with claw tufts. Note that only the claw tuft hairs are shown. *(Redrawn from Kaston 1978.)*

Eyes. Spiders may have from zero to 12 simple eyes that are always paired. Most spiders in Texas have eight eyes, some have six, and rather a few species have less than six eyes (Figure 4). Eyes are key characters in the identification of spider families. The number, arrangement, and size of a spider's eyes are often enough to place the spider in the appropriate family. Eyes are usually in two rows, but they may be in clusters or on tubercles or raised areas. Rows of eyes are described as "procurved" (bent forward at the ends) or "recurved" (bent backward at the ends). In cases where a row of eyes is bent to the extreme, they are counted as two rows.

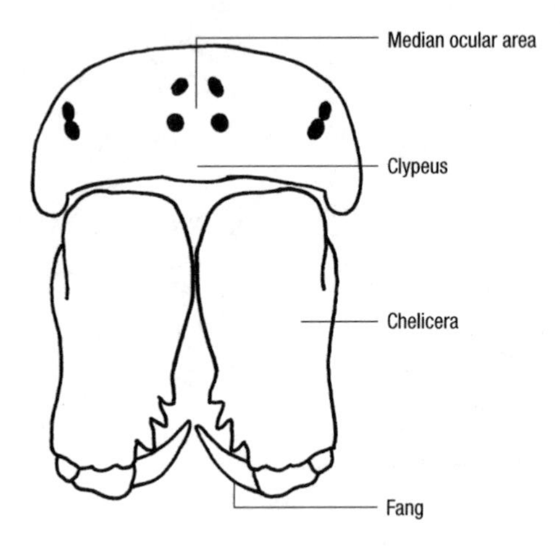

Figure 4. Front of face of a spider. *(Redrawn from Kaston 1978.)*

Chelicerae. The chelicerae are attached at the front of the spider's face (Figure 4). They are usually independent (or free) but may be attached to each other near the base (fused). At the outer tip of the chelicerae are the fangs which inject venom. Spiders feed by first injecting venom to immobilize their prey. Then they secrete digestive enzymes from the gut to dissolve the prey. Spiders ingest only liquid food. Some spider species may manipulate their prey in this process and macerate it or tear it apart with their chelicerae.

Pedipalps. Pedipalps are multifunctional organs that appear to be a small pair of legs near the mouth (Figure 5). They are used to manipulate prey and other objects in the environment.

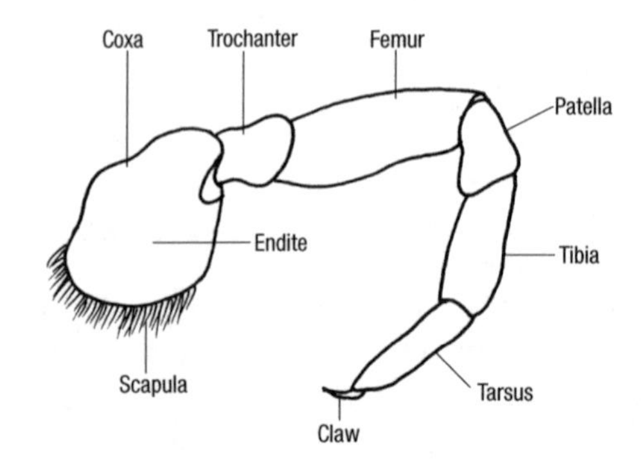

Figure 5. Pedipalp of a female spider. *(Redrawn from Kaston 1978.)*

The most important purpose of the pedipalps is in mating. Female spiders have simple pedipalps that look like small walking legs (Figure 6a). Male spiders have a swollen tip on the pedipalp which is modified for use in mating (Figure 6b). The tip of the male pedipalp consists of small sclerites attached to membranous areas, allowing the sclerites to move. Sclerites are the hardened portion and can be elaborate shapes. The developed pedipalp is the easiest way to sex a spider. A swollen tip of a pedipalp is an indication of a mature male or a penultimate male (Figure 6b). However, females of a few genera and immature males that are ready to molt to adults show slight swelling of the tip of the pedipalp. Male pedipalps have elaborate modifications on their tips, which are inserted into the female epigynum to transfer sperm (Figures 7 and 8). Because the modifications of these sex organs are so elaborate, they are used to identify spiders to species. This requires careful study and a good microscope.

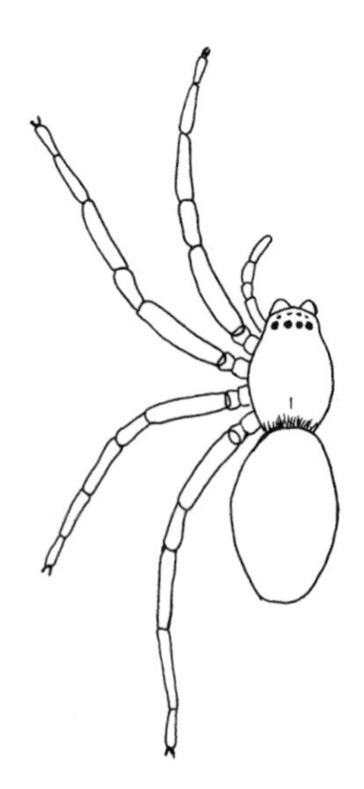

Figure 6a. Female spider.
(Redrawn from Kaston 1978.)

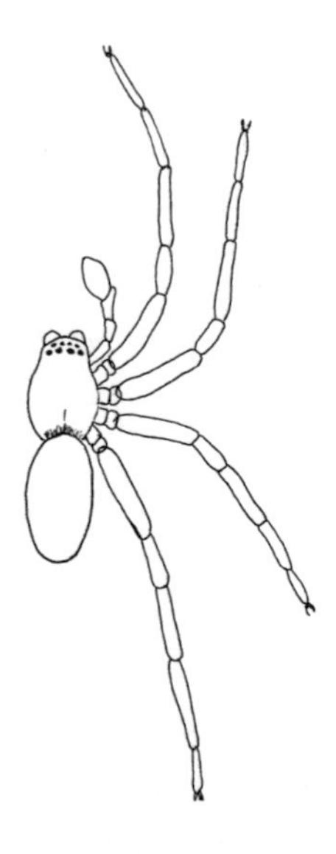

Figure 6b. Male spider.
Note the enlarged tip of the pedipalp, smaller abdomen, and longer legs of male.
(Redrawn from Kaston 1978.)

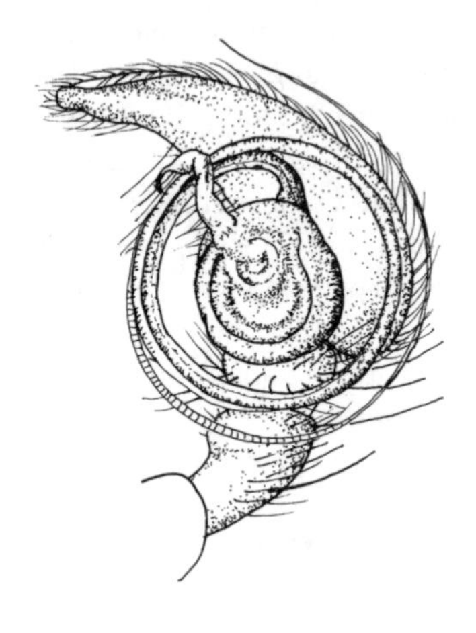

Figure 7. Tip of pedipalp of male *Gnaphosa sericata* shown from below. *(Redrawn from Kaston 1978.)*

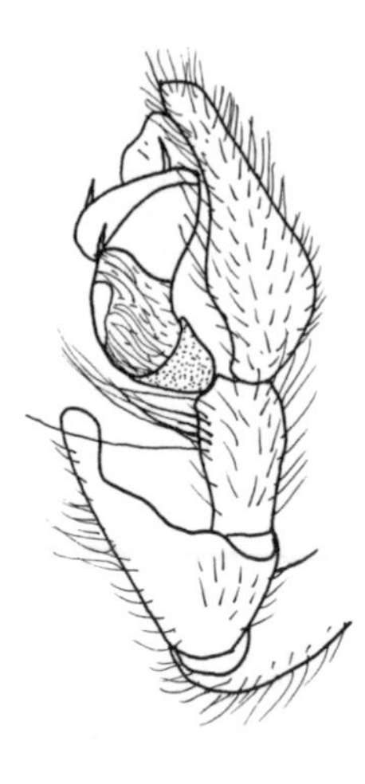

Figure 8. Tip of pedipalp of male *Pityohyphantes costatus* shown from the side. *(Redrawn from Kaston 1978.)*

Underside of the cephalothorax. The underside of the cephalothorax of a spider is formed primarily of one plate called the sternum (Figure 1). The front part of the plate is called the labium. The labium may be separated, forming a separate plate. "Rebordered" labium means that the labium has a thick lip or ridge on the front edge.

Spines and hairs. Spiders have a variety of spines and hairs on their bodies and legs. Spines are stiff extensions of the exoskeletons. Hairs may extend outward from the body or lay flat (decumbent) on the surface. Hairs may be modified into small scoop-like shapes (scopulate hairs) (Figure 9). They may be very dense, forming pads, combs, or other structures. It is common to find more than one type of hair on a spider. The thinnest hairs found on a spider are trichobothria (Figure 10). These may go unnoticed even under high magnification and good lighting. Trichobothria are used for identification of several spider families.

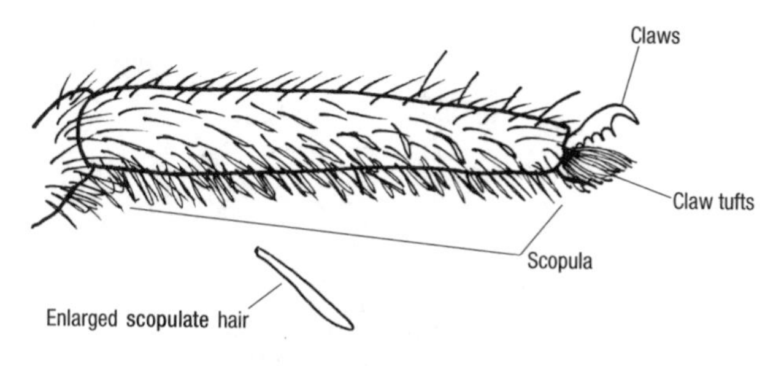

Figure 9. Tarsus of *Tibellus* showing scopula. *(Redrawn from Kaston 1978.)*

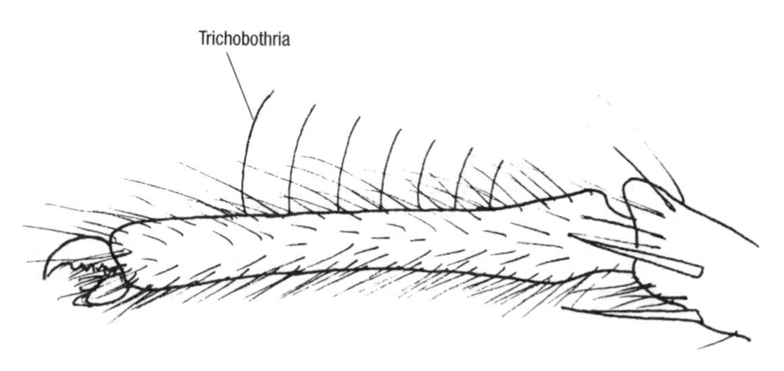

Figure 10. Tarsus of *Coras* showing row of trichobothria. *(Redrawn from Kaston 1978.)*

Abdomen. The spider abdomen can be many different shapes. Spinnerets are small appendages at the end and on the underside of the spider abdomen from which silk is produced (see Figures 1a–1c). There are either six or eight spinnerets, with six the most common. A small tubercle in front of the spinnerets is known as a colulus (the remnants of the first two spinnerets). In some spiders, the colulus is modified to form the cribellum (Figure 11), which is used with a comb-like structure called a calamistrum on the legs (Figure 12) to feather out the silk for certain portions of the web. An anal tubercle may be present behind the spinnerets.

The spider abdomen also has lung slits on the underside (Figure 1) which open internally into book lungs. There is typically one pair of lung slits, toward the front of the spider abdomen. Female spiders also have an epigynum which consists of a pair of slits used in mating to receive sperm from the male. The epigynum may appear like a small plate with two openings, and it may have a single extension called a scape.

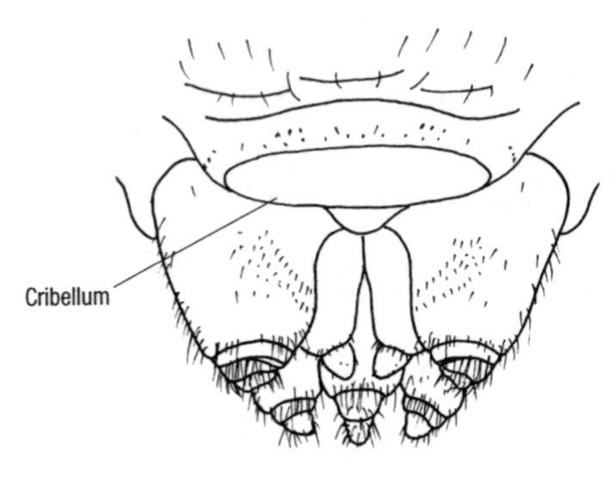

Figure 11. Spinnerets and cribellum of *Hypiotes*. *(Redrawn from Kaston 1978.)*

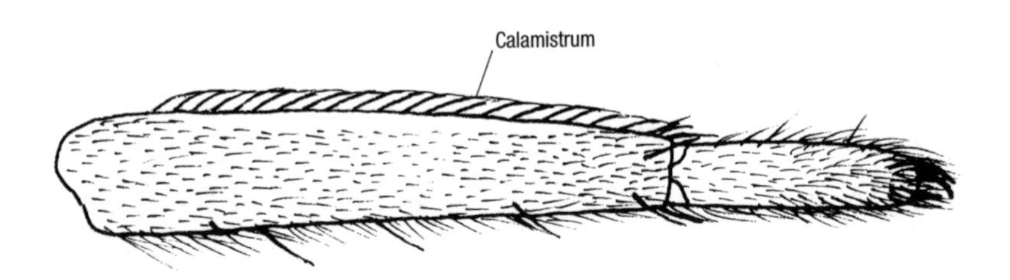

Figure 12. End of the hind leg of *Titanoeca* showing calamistrum on the metatarsus. *(Redrawn from Kaston 1978.)*

Webs and Silk

The terms "web" and "silk" are used almost interchangeably in spider literature to refer to the same material. Generally, silk refers to single strands, while the term web refers to composites used to capture prey or for some other purpose. Each silk strand is really a multistranded chemical fiber which has great strength. Some spiders (the cribellate spiders) can comb out the silk to produce a feathery appearance on the strand. This is apparently an adaptation to produce a stickier silk strand that is more effective at capturing prey.

Silk is used for a variety of purposes by spiders: to form egg cases or eggsacs, as a temporary storage place for sperm, as a shelter or retreat, as drag lines, and as a variety of snares. Eggsacs may be placed on a substrate or hang in a web. Sperm webs are built by some males to receive sperm from their abdomens and hold it until they transfer the sperm to their pedipalps. Shelters are formed like a tunnel or funnel, and may be a separate structure or actually part of the snare. Drag lines are single strands of silk that a spider places down on a substrate as it walks. Many spiders use drag lines to save themselves when they leap into space; they can then pull themselves back up to where they last left. Young spiders or spiderlings also use silk to migrate by "ballooning." Spiderlings move to an open location and release strands of silk to be caught in the air current. After sufficient silk has been extruded to carry the spiderling in the wind, the spiderling simply lets go of the surface. The silk and the attached spiderling will then float, sometimes miles, to a new location.

Snares are used by spiders to capture prey. The most recognized spiderweb is produced by orbweaving spiders with a circular pattern with radii forming the base (Figure 13). The various portions of the web are named and are of some use in identifying spiders. Other spiders form irregular webs which seem to have no plan. Still others produce webs that are more like a sheet or funnel with a retreat at the end. Often, the spiderweb is a clue to the spider family.

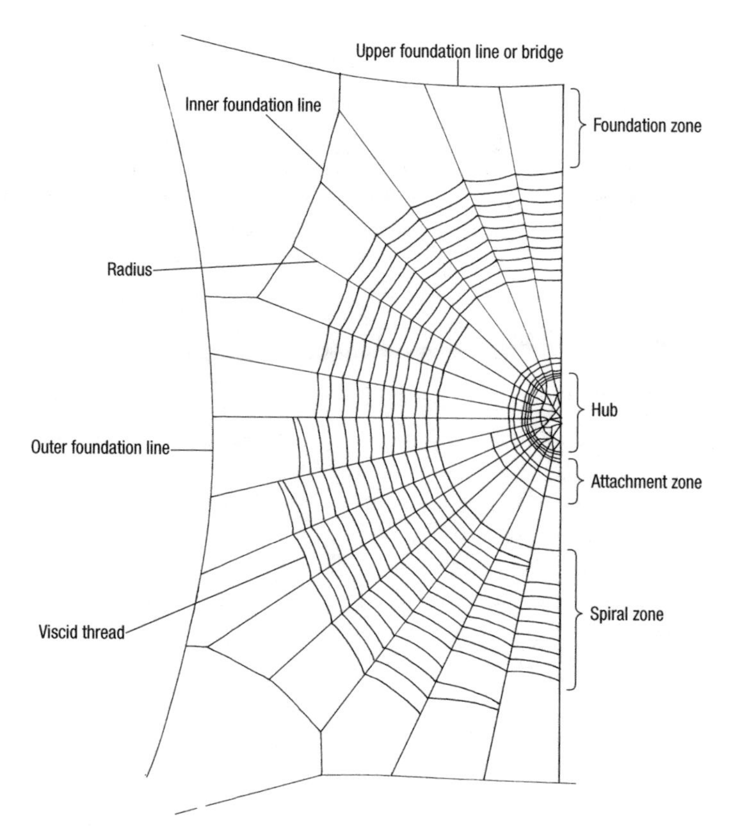

Figure 13. Diagram of the web of an orbweaver spider. *(Redrawn from Kaston 1978.)*

Feeding, Growth, and Reproduction

Feeding. Once prey is captured, feeding can proceed. Spiders must secrete digestive enzymes to dissolve the prey. They only ingest liquids and digestion is external. Prey may be manipulated in the process, and nearly all of the soft body tissues are ingested.

Courtship. Spider courtship is an intricate and sometimes hazardous event. Male spiders must approach a female with caution, because a female may decide the male is food rather than a mate. Males perform a courtship ritual which can be very elaborate in some species. Courtship rituals may include visual displays, tactile displays, vibratory displays, and/or chemicals known as pheromones (Richman and Jackson 1992).

Jumping spider males use a variety of tactics to court females. The male spider transfers sperm from the abdomen to the pedipalp, using a sperm web as a temporary storage site. When a male has reached a receptive female, he will insert the embolus (part of the pedipalp) into the epigynum of the female and transfer sperm. Each pedipalp fits into only one side of the epigynum.

Growth. Spiders have exoskeletons which must be shed for them to grow. Growth occurs by a series of molts. The number of molts varies with species and probably also with food availability and environmental conditions.

Spiders start as eggs, molt into the post-embryo stage, and molt again into the first instar. These first two molts, and sometimes the third, take place inside the eggsac. A typical spider molts 4 to 12 times before it is an adult. The stage before a spider becomes an adult is called the penultimate instar. Some spiders, including mygalomorphs and possibly some araneomorphs, continue to molt even after they have matured. These stages are known as postultimate instars.

Reproduction. The number of offspring produced by spiders may be just a few to several hundred. Eggs are laid in a group, usually in a sac. Some spiders, such as wolf spiders, show maternal care by carrying the eggsac and young spiderlings on their bodies. Most spiders, however, show little attention to their young. Many may even eat them if they linger in the area. Some spiders will tolerate young in the web for short times after they hatch. Relatively few species of spiders are considered social.

Venom

Poisonous Spiders. All spiders have poison glands with the exception of the Uloboridae and some species of Liphistidae. Spiders inject poison when they bite their prey, which helps to immobilize the prey. However, few spiders bite humans even when provoked. Moreover, most spider bites do not cause more than minor discomfort, although individual reactions to spider bites vary. Some sensitive individuals may have a severe reaction to a spider bite which other people would hardly notice. If a reaction includes excessive swelling, loss of body control, difficulty breathing, or other extreme symptoms, seek medical attention at once.

Two groups of spiders, the widow spiders and the recluse spiders, are considered to have venom that is potentially harmful to humans. See the discussions of these species for more information. The next most serious spider bites come from *Cheiracanthium* and can produce slight fever and destroy tissue around the bite in sensitive individuals. *Dysdera* and *Trachelas* may also cause a reaction in some individuals.

Examining Spiders

Examining spiders can be a difficult task. Field notes on color should be taken, or good photographs if possible. Once stored in alcohol, the coloration of a spider changes. Specimens are stored in individual vials in 80% ethanol (ethyl alcohol). Each specimen vial should have collection data on labels that include state, county, distance and direction from nearest town, date collected, and collector. Some information on the habitat should also be included if possible. Labels should be small and neatly written or printed. A separate label with identification information and the authority who identified the specimen should be added to the vial. In general, high-quality India ink or an equivalent works best for long-term storage. Be sure to let the ink dry thoroughly before placing the label in alcohol. A simple #2 pencil will be adequate for temporary labels. Labels printed with a laser printer sometimes flake off unless high-quality paper is used.

A good-quality binocular microscope is almost essential in studying body parts of spiders. Scientists place the specimen in a clear dish that is quite deep, perhaps one inch. The bottom of the dish can be filled with clean sand. Position the specimen directly in the sand to adjust its position for study. Some scientists (Levi 1973) prefer to prepare a dish with paraffin on the bottom with various-sized depressions to hold the specimen. Black paraffin is useful for examining most specimens, but a lighter color may be more useful for dark-colored spiders. Cover the specimen completely with alcohol while studying it under the microscope. Use a high-intensity light. Fiber optic lights that can be moved around are preferred and do not heat up while in use. Often, features can be seen best when the light is from the side or at an angle rather than from above. Direct overhead lighting seldom shows structural features well.

An Overview of Spider Families in Texas

There are 62 families of spiders recorded in North America. Of these, 48 families are recognized in Texas (see table: Spider Families in Texas). These families include the widespread and very common families like Araneidae, Gnaphosidae, Lycosidae, Theridiidae, Thomisidae, and Salticidae. Other families consist of a small number of common species or only a few obscure or localized forms. A summary of these families indicates the diversity within each family and, to some extent, how commonly they might be encountered.

The 14 families that have been found in North America but that have no records in Texas are: Antrodiaetidae (foldingdoor spiders); Mecicobothriidae; Nemesiidae (tubetrapdoor spiders); Hypochilidae (lampshade weavers); Telemidae; Ochyroceratidae; Plectreuridae; Deinopidae (ogrefaced spiders); Theridiosomatidae (ray orbweavers); Symphytognathidae (dwarf orbweavers); Anapidae; Desidae; Zodariidae; and Homalonychidae (dusty desert spiders). It seems likely that several of these families will eventually be found in Texas.

Spiders are divided into two groups or infraorders: the Mygalomorphae and the Araneomorphae. The mygalomorphs are considered the more primitive spiders and are generally large, brown to black, with many living in burrows in the ground. Mygalomorphs always have stout legs and eight eyes, and the cheliceral fangs are parallel. They also have two pairs of book lungs, with the second pair usually visible as white spots posterior to the epigastric furrow. The tarantulas (Theraphosidae) are by far the most common spiders in this group in Texas.

Most of our spiders are in the infraorder Araneomorphae. They come in many sizes, shapes, and colors. Araneomorphs have variable legs and the cheliceral fangs opposing each other or at an oblique angle. Most of them have one pair of book lungs, but some have two pairs. Araneomorphs have twelve or fewer eyes, although most species in Texas have eight or six; relatively few species have less than six eyes.

Some spider families are easy to identify, while others are much more difficult to determine even by an expert. One good method to identify spiders to families is with an identification key. The latest key to spider families in North America is by Roth (1993), but the work by Kaston (1978) still remains very useful if the recent nomenclatural changes are taken into account. Most spider keys require a good microscope, considerable experience, and usually a reference collection before any confidence can be achieved with identifications. Several quick recognition tables for identification are provided in this book rather than a complete key. The reader should be cautioned that these are meant to assist with the majority of specimens and that there are some exceptions. While trying to identify spiders, keep in mind that identification is a process of elimination as much as one of recognition. By eliminating many of the possibilities, a probable identification can often be made.

Spider Families in Texas

(arranged in phylogenetic order*)

Family Name and Common Name	Approx. genera/sp. in Texas	General Range	Abundance in Texas	Habitat	Web Type	Body Length (mm)	Eye No., Pattern	Claws
Infraorder Mygalomorphae								
Atypidae purseweb spiders	1/2	E half of US	rare	ground, in woods	silken tubes horizontal or up tree base	10–25	8	3
Cyrtaucheniidae cyrtaucheniid spiders	2/4	widespread, mostly SW	uncommon	sand dunes, deserts	burrows, some with trap doors	15–23	8	3
Ctenizidae trapdoor spiders	1/5	S USA N to IL	uncommon	river banks, steep slopes	burrows with trap doors	13–26	8	3
Dipluridae funnelweb spiders	1/2	NC, TN, Pacific NW, CA–TX	few records	under rocks, thick vegetation	burrow with funnel web	8–46	8	3
Theraphosidae tarantulas	1/15	SW quarter of USA	common	ground burrows	loose lining of burrows	38–58	8	3
Infraorder Araneomorphae								
Filistatidae crevice weavers	3/4	S USA	common	under stones, crevices, indoors	tubular retreat w/radiating lines	7–19	8 1 cluster	3
Sicariidae sixeyed sicariid spiders	1/5	TN–KS, OK, TX, CA	uncommon	under logs, stones, indoors	none	6–9	6 3 diads	2
Scytodidae spitting spiders	1/5	widespread	common	indoors	none	4–9	6 3 diads	2

After Platnick 1993.

Continued on next page

Spider Families in Texas *(continued)*

Family Name and Common Name	Approx. genera/sp. in Texas	General Range	Abundance in Texas	Habitat	Web Type	Body Length (mm)	Eye No., Pattern	Claws
Leptonetidae cave spiders	2/11	S OR, CA, E AZ, Edwards Plateau of TX, SE USA–Appl. Mtns.	local	caves, forest litter	–	1–2.3	6 or 0	3
Pholcidae daddylongleg spiders	10/18	widespread	common	sheltered areas	irregular, loose sheet	1.6–7	6 or 8 2 triads	3
Diguetidae desertshrub spiders	1/4	SW USA–UT, CO–TX, OK	local	deserts, in low vegetation	irregular, vertical tubular retreats	5–6	6 3 diads	3
Caponiidae caponiid spiders	2/2	S CA, NV–TX	uncommon	under stones, arid regions	–	–	2 or 8	3
Segestriidae segestriid spiders	1/1	E US	uncommon	tree cracks, under stones	tubular retreat	5.5–15	6	3
Dysderidae dysderid spiders	1/1	widespread	uncommon	under stones, bark, moss	flat oval retreat	9–15	6 transverse oval	2
Oonopidae oonopid spiders	4/9	mainly FL, NE USA, S states N to UT	uncommon	grass, bushes, indoors	unknown	<3	6	2
Mimetidae pirate spiders	2/7	widespread	uncommon	bushes, stones, spiderwebs	no snares	2.3–5.6	8	3
Oecobiidae flatmesh weavers	1/3	widespread	common	under stones, window ledges	flat mesh	2–3	8	3

Hersiliidae longspinneret spiders	1/1	S TX, FL	local	tree bark or hollows	retreat, loose sheet	12	8	3
Uloboridae hackled orbweavers	4/7	widespread	common	foliage	orbs, sectors of orbs	2–6	8 2 rows	3
Nesticidae cave cobweb spiders	2/7	widespread	rare	caves, under stones	cobweb	3–4	8	3
Theridiidae cobweb weavers	22/87	widespread	very common	foliage, indoors, anywhere	irregular cobweb	1–13	8	3
Mysmenidae dwarf cobweb weavers	1/1	E of Miss. R., S TX, AZ, OR, UT	rare	–	orb	<2	8	3
Linyphiidae sheetweb and dwarf weavers	22/49	widespread	very common	foliage, leaf litter	platforms, domes	1.6–8.5	8	3
Tetragnathidae longjawed orbweavers	7/16	widespread	common	mostly foliage, often near water	orbs, often at an angle	4–9	8	3
Araneidae orbweavers	27/84	widespread	very common	wooded areas, foliage	orbs	2–28	8	3
Lycosidae wolf spiders	15/63	widespread	very common	forest floor, grasses, shores	lined burrows	3–35	8 3 rows, 2nd pair large	3
Pisauridae nursery web spiders	3/8	widespread	common	near water, bushes, grass	nurseries	9–28	8 2 rows	3
Agelenidae funnel weavers	4/11	widespread	very common	grass, under leaves, stones, in bushes	funnel	3–20	8 few with 6	3
Hahniidae hahniid spiders	2/6	widespread	uncommon	moist places near ground	delicate sheets	1.5–2.6	8	3

Continued on next page

Spider Families in Texas *(continued)*

Family Name and Common Name	Approx. genera/sp. in Texas	General Range	Abundance in Texas	Habitat	Web Type	Body Length (mm)	Eye No., Pattern	Claws
Dictynidae meshweb weavers	11/107	widespread	common	tops of weeds, twigs, under stones, on ground	irregular snares	1.3–2.2	6–8	3
Amaurobiidae hackledmesh weavers	2/3	widespread	common	under stones, leaf litter, bark	radiating snare with central retreat	3.5–12	8	3
Titanoecidae titanoecid spiders	1/2	widespread	uncommon	loose stones, dead leaves	loose hackled web	3.5–8	8	3
Tengellidae tengellid spiders	1/3	TX–AZ	rare	vertical sides of bare soil banks, under rocks	open tubular burrows	–	8	2
Oxyopidae lynx spiders	3/12	widespread	very common	low bushes, vegetation	no snares, retreats	3.4–16	8 6 in hexagon, 2 in front	3
Anyphaenidae ghost spiders	5/17	widespread	common	foliage, under logs, forest floor	no snare	3–8.4	8	2
Miturgidae miturgid spiders	3/4	widespread	uncommon	under stones	silken retreat	7–9	8 2 rows	2
Liocranidae liocranid spiders	3/8	widespread	common	under leaves, rocks	no snare	2–13	8 2 rows	2
Clubionidae sac spiders	3/13	widespread	very common	foliage, under bark	no snare, retreats	3–11	8	2

Corinnidae antmimic spiders	5/18	widespread	common	under bark, rolled leaves, indoors	no snare	3–10	8	2
Prodidomidae prodidomid spiders	1/1	S USA	rare	–	–	–	8	2
Gnaphosidae ground spiders	20/90	widespread	very common	under stones, logs, in pastures	tubular retreats	3–15	8 2 rows	2
Ctenidae wandering spiders	3/3	SE USA–SW TX	uncommon	foliage, ground	–	6–18	8 3 rows with 2, 4, and 2	2
Zoridae zorid spiders	1/1	E Coast W to AL, TX, OR, CA	rare	tall grass, bushes	no snare, no retreat	3.5–6	8 3 rows 4 in front	2
Selenopidae selenopid crab spiders	1/1	S CA–S TX, FL	rare	under stones	–	9–12	8, 6 large in front	2
Heteropodidae giant crab spiders	2/2	extreme SW USA, S tip of FL	local	–	–	10–23	8 2 rows	2
Philodromidae running crab spiders	5/33	widespread	common	bushes, trees, grasses	no snares, retreats	2-10	8 2 rows	2
Thomisidae crab spiders	9/41	widespread	very common	bushes, trees, flowers	no snares, retreats	2–11	8 2 rows	2
Salticidae jumping spiders	45/114	widespread	very common	bushes, trees, grass, indoors	no snares, retreats	2–15	8, 3–4 rows 2 large eyes in center front	2

Quick Recognition Guide to Some Common Spider Families in Texas Based on Eye Patterns

Characters	Family
Spiders with no eyes or only eye spots	Dictynidae (*Cicurina*–in part) Linyphiidae (*Islandiana*–in part) Nesticidae (*Eidmannella*–in part)
Spiders with four eyes	Uloboridae (*Miagrammopes*)
Spiders with six eyes	
Eyes in one cluster	Filistatidae Pholcidae (in part)
Eyes in three pairs (diads) (Figure 14)	Diguetidae Scytodidae Sicariidae
Eyes in two triads (Figure 14)	Pholcidae (in part)
Eyes otherwise	Agelenidae (in part) Dictynidae (in part) Diguetidae Dysderidae Linyphiidae (in part) Nesticidae (in part) Segestriidae
Spiders with eight eyes	
Eyes in one cluster	Atypidae Ctenizidae Cyrtaucheniidae Dipluridae Theraphosidae
Three or four rows with anterior medians large (Figure 14)	Salticidae
Two largest eyes in hind row, front row of four similar in size (Figure 14)	Lycosidae
Two eyes in front, remaining six form a hexagon (Figure 14)	Oxyopidae
Four large eyes in hind row, both rows recurved	Ctenidae
A pair of eyes in front plus two triads (Figure 14)	Pholcidae (in part)
Two rows of four	Agelenidae (in part) Amaurobiidae Anyphaenidae Araneidae Clubionidae Corinnidae Dictynidae (in part) Gnaphosidae Hahniidae Hersiliidae

Characters	Family
Two rows of four	Heteropodidae Linyphiidae (in part) Liocranidae Mimetidae Miturgidae Nesticidae (in part) Oecobiidae Philodromidae Pisauridae Selenopidae Tetragnathidae Theridiidae Thomisidae Titanoecidae Uloboridae (in part)

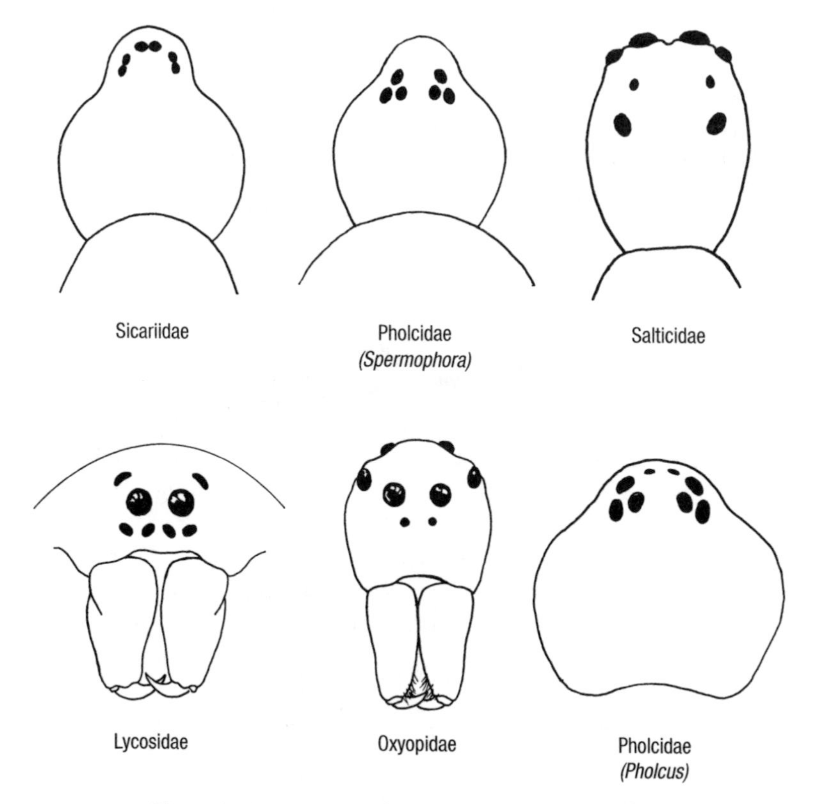

Figure 14. Eye patterns of some common spider families.

Quick Recognition Guide to Some Common Spider Families in Texas Based on Web Type

Web type	Family
Orb webs or sectors of orbs	Araneidae Tetragnathidae Uloboridae
Cobweb or irregular webs	Amaurobiidae Dictynidae Nesticidae Pisauridae Theridiidae Titanoecidae
Funnels, sheet-like platforms, retreats with radiating threads	Agelenidae Dipluridae Filistatidae Hahniidae Hersiliidae Linyphiidae Oecobiidae Pholcidae
Tubes and burrows (often in or near the ground)	Atypidae Ctenizidae Cyrtaucheniidae Lycosidae (in part) Theraphosidae
No prey webs (only retreats)	Anyphaenidae Clubionidae Corinnidae Dysderidae Gnaphosidae Liocranidae Lycosidae (in part) Mimetidae (in webs of other spiders) Miturgidae Oxyopidae Philodromidae Salticidae Scytodidae Segestriidae Sicariidae Thomisidae

Quick Recogniton Guide to Spider Families in Texas Based on How They Carry Eggsacs

Structure holding eggsac	Family
Chelicerae	Pholcidae Scytodidae
Chelicerae and palpi under sternum	Pisauridae
Spinnerets	Lycosidae (most)
Spinnerets and one hind leg	Nesticidae (in part) Theridiidae (in part)

INFRAORDER MYGALOMORPHAE

1 ATYPIDAE—PURSEWEB SPIDERS

Key Family Characters

Purseweb spiders are typically brown to black, relatively large (10–30 mm), and have the chelicerae projecting forward as is typical of the mygalomorphs. They have six spinnerets, abdominal tergites, and a transverse thoracic depression. The absence of a depression or groove between the labium and sternum is a key character that defines the family (Roth 1993). The purseweb tube of silk that extends from the ground is characteristic.

Biology

These spiders form a purseweb, or tube of silk, in which they hide. They capture prey that land on the outside of the tube by biting the prey through the tube. The spider cuts a slit in the tube and pulls the prey inside to feed on it.

The silken tubes are either horizontal on the ground or extending up the bases of trees for about 8 to 10 inches.

Taxonomic Status

The only genus in this family recorded from Texas is *Sphodros.* There are only a few records from the state. *S. paisano* Gertsch & Platnick is recorded from Cameron and Travis counties; *S. rufipes* (Latreille) is reported from Liberty County.

References

Levi et al. (1990). Kaston (1978). Roth (1993). Gertsch (1979). Gertsch and Platnick (1980).

2 CYRTAUCHENIIDAE—CYRTAUCHENIID SPIDERS

Key Family Characters

Spiders in this family are mygalomorphs which lack abdominal tergites and claw tufts. They are easily confused with the Ctenizidae, and both form burrows in the ground. See Roth (1993) for definitive characters that separate these families.

These are large spiders, up to 25 mm in length.

Biology

These spiders form burrows usually closed with a trap door (Roth 1993). Webs are used to line burrows and form a hinge for the trap door.

Taxonomic Status

This family has been included in the Ctenizidae in the past. Genera that occur in Texas are *Eucteniza* and *Myrmekiaphila*.

Reference

Roth (1993).

3 CTENIZIDAE—TRAPDOOR SPIDERS

Key Family Characters

Spiders in this family are mygalomorphs which lack abdominal tergites and claw tufts. They are easily confused with the Cyrtaucheniidae, and both form burrows in the ground. The only genus recorded from this family in Texas is *Ummidia* which can be recognized by a saddle-shaped bend in the tibia of the third leg. See Roth (1993) for definitive characters that separate these families.

These spiders are primarily found in southern and western states. The family is widespread and is probably more common than the few records would indicate.

These are large spiders, up to 25 mm in length.

Biology

These spiders have the habit of digging tubular burrows in the ground. The walls of the burrow are lined with silk, and the entrance is covered with a solid lid (trap door) with a silk hinge. The lid fits snugly and may be further camouflaged with bits of vegetation. The spider may hold the door closed tightly. Doors may be thin "wafer" types or thicker "cork" types. Some species build tunnels with a branch somewhat like an inverted Y, while others are more L-shaped.

Taxonomic Status

In the past, the family has included genera that are now included in Cyrtaucheniidae.

Reference

Kaston (1978).

4 DIPLURIDAE—FUNNELWEB SPIDERS

Key Family Characters

These mygalomorph spiders have extremely long spinnerets, either four or six, with the longest spinnerets half the length of the abdomen or longer. They are small to large spiders with lengths of 2.5 to 18 mm.

Biology

These spiders produce funnel-shaped webs to catch prey, and one corner may be used as a retreat. Funnels may be at the bases of trees, among rocks or wood, or in thick vegetation.

At least some species extend the funnels into large sheets over the ground.

Taxonomic Status

This is a small family with three genera and only a few species, primarily in the west and southwest. *Euagrus* is the only genus known to occur in Texas.

Euagrus comstocki Gertsch are medium to large spiders (over 6 mm), the thoracic furrow is a circular pit, chelicera lack a rastellum, tarsi lack scopulate hairs, and the long spinnerets are separated by at least the diameter of a spinneret.

The cephalothorax and legs are brown, the abdomen tan to dirty yellow. The cephalothorax is flat above with the head region not higher than the thoracic region. Anterior spinnerets are separated by three times the diameter of one eye.

Length of the female 15 mm; length of the male 13 mm.

Range

Records for this species are primarily from south Texas. Another species in the genus, *E. chisoseus* Gertsch, is found in central and west Texas.

Outdated and Unofficial Names

Evagrus.

References

Kaston (1978). Levi et al. (1990). Coyle (1988). Roth (1993).

5 THERAPHOSIDAE—TARANTULAS

Key Family Characters

Tarantulas are our heaviest spiders by weight. They are relatively common throughout the state and their large size makes them quite recognizable.

The tarantulas can be separated from other mygalomorphs by the claw tufts on the tarsi and the presence of a patch of urticating hairs on the abdominal dorsum. (Roth 1993). They range in size to 40 mm or larger.

Biology

Tarantulas use burrows in natural cavities under logs or stones or under loose bark of tree trunks. They sometimes take over old rodent burrows but also dig their own burrows. These spiders are usually restricted to the ground but can climb.

Mating is initiated when the male spider approaches the female with his front legs raised. He catches her fangs on the spurs of his front legs and bends her backwards. The male then pushes the embolus of the palpal bulb into the epigastric furrow of the female and discharges his sperm. He repeats the process with the other palpal bulb. There is much swaying and tapping of the palpal bulb on the underside of the female in the process, which may calm the female during mating.

Webbing is sometimes used to line the tarantula's shelter, and a few lines of silk are placed on the ground in front of the shelter. In some cases, the silk lines actually form a mat. The silk threads transmit vibrations from passing prey to the spider in the shelter.

Males build sperm webs to temporarily hold their sperm. They deposit a drop of sperm usually on the top of the web from the genital opening of their abdomen. They then turn around and charge the palpi with sperm. After completing this operation, the male usually destroys the sperm web.

Tarantulas are often kept as pets. They calm quickly in captivity and may become accustomed to handling. They can be directed into a cup to move them. Some handlers put their hands into the containers and let the spiders crawl onto them. Experienced handlers pick tarantulas up by the sides of the carapace. Some tarantulas move so fast that handling is difficult. They may be fed wild foods that are collected or commercially available live foods such as mealworms, crickets, and even baby mice. They also need a supply of water that they can place their mouth into, so a shallow dish of water should be provided. A number of species in this family are now imported in the pet trade, and books on care and maintenance are available.

Taxonomic Status

A recent study by Smith (1994) attempts to update the taxonomy of this family.

References

Kaston (1978). Levi et al. (1990).

5A. *Aphonopelma*

Description. Typically, the cephalothorax and legs are dark brown and the abdomen brownish black. Color may vary between individuals and certainly changes after a molt. These are our largest spiders by weight.

Length about 40 mm.

Biology. Females lay 100–1,000 eggs in a web which is constructed like a hammock. The silk is spun over the eggs after they are laid to form an eggsac. The eggsac is retained in the burrow, guarded, and usually held by the female. Eggs hatch in 45–60 days. Spiderlings hatch in July or later in the year within the eggsac. Once they leave the eggsac, the spiderlings may stay with the females for 3–6 days or longer before dispersing. Many of the young fall prey to other spiders or predators as they disperse to begin their own burrows. Females have lived in captivity for over 25 years. Males in Texas rarely live over two or three months after maturity.

They usually remain in their burrows waiting for prey to come by, but may move a few meters out to forage when necessary. They typically feed on crickets, June beetles, ground beetles, grasshoppers, cicadas, and caterpillars.

When disturbed, tarantulas maneuver to face the threat and will raise up on their hind legs and stretch out their front legs in a threatening posture. When disturbed, they also may rapidly brush the top of their abdomens with their hind legs which dislodges urticating hairs. These hairs can irritate the eyes or skin of a vertebrate attacker.

Wasps in the genus *Pepsis* attack tarantulas and paralyze them with a sting. They then drag them to a burrow in the sand where the spider is used as food on which the young wasps develop. There are many variations in the way these wasps complete this procedure.

A recent study (Janowski-Bell and Horner 1995) used radio tracking techniques to monitor activity in September and October. Tarantulas were active from about two hours before sunset to two hours after sunrise. Mostly males were found away from their burrows. One female monitored away from her burrow was in the process of mating.

One of the most spectacular spider events in Texas occurs for a few weeks each summer when male tarantulas actively wander, apparently seeking females. Spiders can be commonly seen crossing the road in the evenings for a few weeks in the summer in central Texas. Oddly enough, most of the spiders seem to be moving east or northeast. This phenomenon is not well understood and may be related to migration more than mating.

Their large size and hairiness often attract attention and concern. Bites of Texas species are generally not serious to humans.

Range. Throughout Texas.

Habitat. Common in grasslands and semi-open areas, but possible nearly anywhere.

Taxonomic Status. There are 14 species in *Aphonopelma* listed from Texas in a recent work (Smith 1994). Identification of species is difficult and requires mature males, a microscope, proper literature, and experience. The large size of these spiders demands larger collection and storage equipment than most spider collectors utilize.

References. Kaston (1978). Levi et al. (1990). Smith (1994). Janowski-Bell and Horner (1995).

Infraorder Araneomorphae

6 Filistatidae—Crevice Weavers

Key Family Characters

The oval shape of the carapace is a distinguishing feature of the family. The eight eyes in a cluster distinguish them from some spiders that look similar. The cribellum may be difficult to see. Males have much longer legs than females and are generally lighter in color and thinner. Females lack an epigynum.

Biology

Crevice weavers hide during the day in tubular retreats that line cracks or crevices. They produce a few strands of silk that radiate from their retreats to sense prey as it walks by or alights on the webs.

Filistatids build snares under stones or around cracks and crevices. They are likely to be found in or around abandoned buildings.

Taxonomic Status

This is a small family with only one species, *Kukulcania hibernalis* (Hentz), frequently encountered.

References

Levi et al. (1990). Kaston (1978). Roth (1993). Lehtinen (1967).

6a. *Kukulcania hibernalis* (Hentz), Southern House Spider

Description. The eyes are close together on a raised prominence. The tracheal spiracle is in front of the spinnerets almost one-third the distance to the epigastric furrow. Males are yellowish tan in color with a darker brown stripe from the eyes to the thoracic groove. They have the femur and tibia about equal in length, and each slightly longer than the carapace. Males also have distinctive long, spindly pedipalps that project forward.

Females look very different than males and are uniformly dark brown or have some irregular dusky blotches on the carapace. The abdomen has black pubescence.

Length of the female 13 to 19 mm; length of the male 9 to 10 mm.

Biology. This spider constructs a tubular retreat usually in a crack or hollow in a wall or corner. It constructs web threads radiating from the retreat on the surface. The radiating threads are quite conspicuous, especially after they have accumulated some dust. The webs may resemble the web of a funnel weaver. The radiating threads send vibrations from passing insects to the spider in the retreat. The spider then runs out after the potential victim. Normally, they come out only at night when attacking prey.

Mature males can be seen out in the open in buildings and around outside walls when they are seeking females. The males usually die within a few weeks of mating. They are commonly found in bath tubs.

Females lay about 200 eggs in a loosely swathed silken ball about 15 mm in diameter. Females may live up to eight years. All life stages may be present throughout the year.

Males especially, but sometimes females and immatures, are confused with recluse spiders. However, filistatids have eight eyes close together while recluse spiders have six eyes in three pairs. There is also no trace of the violin marking on the filistatid, but they may have a darkened stripe on the carapace. They are not known to bite unless strongly provoked.

Range. Southern states west to California.

Habitat. Retreats are usually in cracks or crevices inside or around the outside of houses, barns, or outbuildings. Occasionally these spiders are found in cotton fields in the eastern half of Texas where they build webs from a crack in the soil or from under a stone, usually on or near the ground.

Outdated and Unofficial Names. *Filistata hibernalis* Hentz, crevice spider.

References. Kaston (1978). Levi et al. (1990). Breene et al. (1993b). Platnick (1989, 1993). Comstock (1940). Edwards (1983).

7 SICARIIDAE—SIXEYED SICARIID SPIDERS

Key Family Characters

Sicariid spiders have six eyes in three pairs or diads and a carapace that is quite flat when viewed from the side and highest near the head (Figure 15). The chelicerae are fused at the base. The shape of the carapace and darkened violin markings are also useful identification marks for our most common species.

Biology

These spiders can be found under logs, stones, or other sheltered areas. They prefer undisturbed habitats and can be found indoors where they hide in dark corners, in trunks, under stored clothing, and around almost any undisturbed structure. They are commonly seen moving around at night, undoubtedly hunting.

The brown recluse, *Loxosceles reclusa* Gertsch & Mulaik, is the most well-known spider in the family because it has venom that is potentially harmful to humans. However, the other members of the genus also have similar venom. Recluse bites usually form a red circular area on the skin which sloughs off, leaving an open wound which is difficult to heal and may require several months. Seek medical attention if such a bite occurs. Of course, severe reactions that require immediate attention are also possible depending on the sensitivity of the individual.

Sicariids form loose, irregular sheet webs which are used to capture prey and as retreats. They also form loosely constructed eggsacs that are attached to surfaces.

Taxonomic Status

This family was formerly placed in the Scytodidae. Moved to Sicariidae (Platnick 1993).

Outdated and Unofficial Names

Loxoscelidae, recluse spiders, violin spiders, brown spiders.

References

Kaston (1978). Levi et al. (1990). Gertsch and Ennik (1983).

7A. *Loxosceles reclusa* Gertsch & Mulaik, brown recluse

Description. The color is generally yellowish brown but can vary. The most distinctive mark is the darker violin shape on top of the carapace. The base of the violin mark is at the front of the carapace, and the "neck" of the violin extends backward toward the abdomen. Females have the first leg almost 4.5 times as long as the carapace, with the femur of that leg slightly longer than the carapace. Males have the first leg about 5.5 times the length of the carapace, and the femur alone is 1.67 times as long as the carapace. Measurements using outstretched legs suggest it is bigger than it is.

Length of the female 9 mm; length of the male 8 mm.

Biology. Mating season is from April to July. Females produce up to 5 eggsacs containing about 50 eggs each. The offspring take nearly one year to develop into adults. Laboratory records indicate that they may live for several years.

L. reclusa is the species most frequently associated with bites of medical significance. This spider is quite nonaggressive, and bites occur most frequently when it is injured or trapped in clothing or bedding. Most bites occur on the buttocks or legs, and typically produce local pain and itching which may take days or over a week to be noticed. Systemic (generalized internal) reactions usually are evident within 72 hours. The bite site may develop a discolored pustulate area that progresses to a necrotic area with an open wound the size of a quarter or larger. Systemic reactions may include: rashes, fever, generalized itching, vomiting, diarrhea, shock, or death. Bites have been treated by surgical removal of the site (preferably

Figure 15. Carapace of *Loxosceles* (Sicariidae) showing general shape and eye arrangement.

early), steroids, oral dapsone, ice packs, aspirin, oral antibiotics, and some antivenims (Demmler et al. 1989). Nitroglycerin patches are also used. Surgical removal of the bite site is not generally advised.

Other species in the genus should also be considered poisonous. Medically significant bites have been reported from: *L. deserta* Gertsch; *L. laeta* (Nicolet), chilean recluse; and *L. rufescens* (Dufour), mediterranean recluse. The venom of *L. reclusa* seems to be stronger than that of *deserta*.

Range. Throughout Texas (although there are no records yet from the Lower Rio Grande Valley). Reported from Ohio south to Georgia and west to Nebraska and Texas.

Habitat. Under bark, under stones, in log piles, and indoors.

Related Species. Gertsch and Ennik (1983) recognize thirteen species in North America. The other species in the genus recorded from Texas are: *apachea* (Gertsch & Ennik); *blanda* (Gertsch & Ennik), the Big Bend recluse, from West Texas; *devia* (Gertsch & Mulaik) from Central and South Texas; and *rufescens* from scattered locations across the state. *L. devia,* the Texas recluse, may be the most common species in Texas and is often mistaken for *L. reclusa.* The larger dark South American species, *L. laeta,* has been accidentally introduced into the United States but is found mostly in southern California. The possibility of it occurring in Texas always exists. It is reported to have the most virulent venom of any of the *Loxosceles* tested (Kaston 1978).

These other species are very similar, but the "violin" markings vary somewhat. Species determination should be completed using genitalia. Relative length of the first leg compared to the size of the carapace has been suggested by some authors for species determination but is not very reliable.

References. Kaston (1978). Demmler et al. (1989). Gertsch and Ennik (1983) have produced the latest revision of the genus.

8 Scytodidae—Spitting Spiders

Key Family Characters

Spitting spiders have six eyes in three diads–a character shared by sicariids and diguetids. Spitting spiders have nearly rounded carapaces when viewed from the top. They have a sloped shape to the carapace when viewed from the side, with the highest portion near the rear. This combination of characters is usually sufficient to distinguish them from other families of spiders. They are spindly-legged and slow-moving spiders.

Biology

Spitting spiders are cosmopolitan in buildings. They hide during the day and move around at night. They squirt a venom-mucus substance that can pin down and kill or paralyze prey without injection.

The poison glands are enormously developed in this family, and each venom gland has two parts. The small front part of the poison gland produces venom which can be injected. The larger posterior portion of the gland produces the mucilaginous substance used to spit on and thus capture the prey. The cephalothorax is very large and elevated to accommodate the large venom glands.

Females carry eggsacs in their jaws.

Some spitting spiders may build snares, but they are generally not recognized as snare-producing. They may use some webbing to line the retreat used during the day.

Taxonomic Status

Kaston (1978) indicates that there are seven species in the United States. The species in Texas (or anywhere else) are not well known. One common species remains undescribed.

All of our species are in the genus *Scytodes.* The general color is yellowish to gray. Spitting spiders often have a mottled coloration, at least on the cephalothorax, and spindly, banded legs. Members of this genus are quite common walking about slowly in shaded corners, windows, dark closets, and similar places. They may be found in almost any corner of a building or storage area. They can be found throughout Texas. Length of the female is 4 to 5.5 mm; length of the male is 3.5 to 4 mm.

References

Kaston (1978). Levi et al. (1990). Valerio (1981).

PHOLCIDAE—DADDYLONGLEG SPIDERS

Key Family Characters

Most pholcids have eight eyes, but some have six (Figures 16 and 17). Eyes are in clusters, and the shape of the carapace and eye positions are distinctive for each genus. The anterior median eyes are the smallest (or entirely lacking), while the other eyes are arranged in two triads. The legs are extremely long and thin and the tarsi flexible.

These long-legged spiders may sometimes be confused with another group of arachnids sometimes called daddylonglegs (Opiliones or harvestmen) because of their long, thin legs. Harvestmen have only one body region, do not build webs, and do not produce venom.

Biology

Male spiders in this family have large, rather simple palps. The female may carry a round eggsac in her chelicerae or eggs may be laid in the web held in place by only a few fibers. Males and females may be found together. *Physocyclus* seems to prefer spiders as prey.

Daddylongleg spiders produce an irregular web in which they hang upside down. The snares are either sheetlike or irregular. Many live in cellars or dark places, while others live under stones, under ledges, or in caves.

Taxonomic Status

About 36 species (Roth 1993) occur in North America, and about 200 worldwide. There are 10 genera and 18 species reported in Texas.

Outdated and Unofficial Names

Cellar spiders, squinteyed spiders.

References

Kaston (1978). Gertsch and Mulaik (1940). Gertsch (1982).

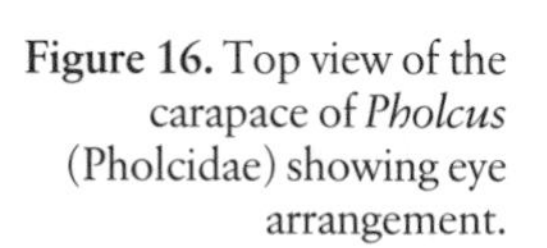

Figure 16. Top view of the carapace of *Pholcus* (Pholcidae) showing eye arrangement.

Figure 17. Carapace of *Spermophora* (Pholcidae) showing shape and eye arrangement.

Generic Summary for Pholcidae in Texas

Quick recognition characters	Genus	Number of species
Eyes clustered on a single tubercle	*Modisimus*	1
Abdomen elongate	*Pholcus*	1
	Smeringopus	1
Smallest species 1.5–3.0 mm with PT/C under 200	*Pholcophora*	2
Six eyes, pale yellow eyes; carapace with black spots	*Spermophora*	1
Other genera not easily recognized		
	Crossopriza	1
	Metagonia	1
	Micropholcus	1
	Physocyclus	3
	Psilochorus	6

9A. *Pholcus phalangioides* (FUESSLIN), LONGBODIED CELLAR SPIDER

Description. The color is pale yellow except for a gray mark in the center of the carapace. There are eight eyes, with two small eyes in front of two triads of larger eyes. This spider has very long, spindly legs (Figure 18).

Length of the female 7 to 8 mm; length of the male 6 mm.

Biology. This spider is found throughout the world and is the most common cellar spider throughout the United States. When disturbed, this spider vibrates the web so fast that it becomes a blur.

Range. Records from Texas include Travis and Wichita counties; however, it is undoubtedly more widespread.

Habitat. Typically found in basements, houses, or dark undisturbed locations.

References. Kaston (1978). Levi et al. (1990).

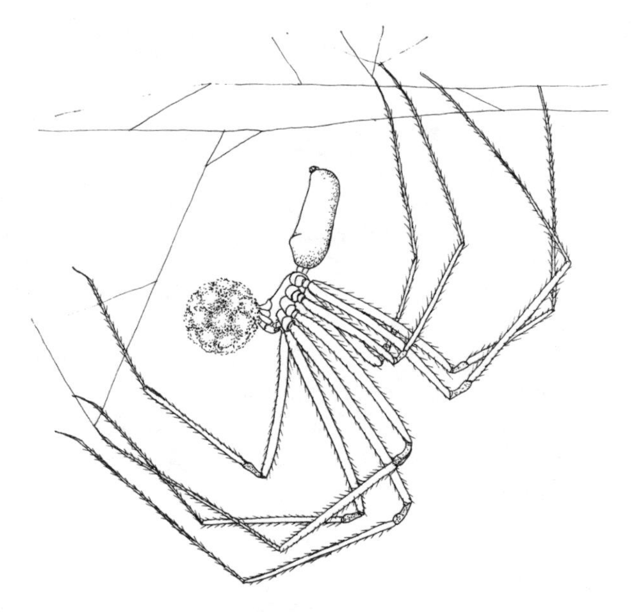

Figure 18. Female *Pholcus* holding eggsac in her chelicerae. *(Redrawn from Kaston 1978.)*

10 DIGUETIDAE—DESERTSHRUB SPIDERS

Key Family Characters

Diguetids share the characteristic pattern of six eyes in three diads with the sicariid and scytodid spiders. The cephalothorax is relatively long and the abdomen is hairy. The few species known are from the southwestern states, Mexico, and Argentina.

Biology

This spider is usually found in webs in desert locations, often within cacti and desert shrubs. It makes a vertical silk tubular retreat covered with eggsacs and remains of prey over the center of an irregular web. At the center of this web is a vertical tubular retreat closed at the top. The eggsacs are placed within the tube.

Taxonomic Status

This is a small family with only one genus in Texas. *Diguetia canities* (McCook) has an orange carapace with white pubescence. The abdomen is tan with a white-bordered folium on the dorsum and is covered with white pubescence. Legs are yellow, except orange-brown on patellae and distal ends of other segments, and are more deeply pigmented in the male than female. Length of the female is 8 to 9.5 mm; length of the male is 5.6 to 6.2 mm.

This spider builds tubes inside the web. It is found primarily in desert areas around shrubs and cacti. It has been recorded primarily in south and west Texas. It is also known from Oklahoma and Texas west to California.

This species has a subspecies *mulaiki* Gertsch that also occurs in south and west Texas. There are two other species in the genus that occur in the western portions of the state. *D. albolineata* O.P.-Cambridge is similar in general appearance to *canities,* but smaller and with proportionally longer legs. Length of the female 6.3 mm; length of the male 5.1 mm.

References

Kaston (1978). Levi et al. (1990). Gertsch (1958).

11 SEGESTRIIDAE—SEGESTRIID SPIDERS

Key Family Characters

These spiders have six eyes closely grouped and a long labium. They have three pairs of legs that point forward, which distinguishes them from spiders with six eyes and three claws (Roth 1993).

Biology

Usually seen at night at the entrance of silken tubes on soil banks or rock walls. All segestriids are nocturnal and capture prey in web tunnels with silk strands radiating from the centers.

Taxonomic Status

Only one genus occurs in North America, and only one wide-ranging species has been reported from Texas. Separated from the Dysderidae by Forster and Platnick (1985).

Ariadna bicolor (Hentz) has the cephalothorax brown, relatively long, and narrow. The abdomen is long and oval, mostly purplish brown above and below, paler at the sides. Length of the female 8 to 10 mm; length of the male 7 mm.

The spider builds a tubular retreat in cracks of trees, under rocks, and under bark. It is found primarily in the eastern two-thirds of Texas, New England south to Florida, and west along the central and southern states to Colorado, New Mexico, and California.

Outdated and Unofficial Names

Dysderidae, *Ariadne.*

References

Roth (1993). Beatty (1970). Kaston (1978). Levi et al. (1990).

12 DYSDERIDAE—DYSDERID SPIDERS

Key Family Characters

The very large, projecting chelicerae, with almost parallel fangs and a deeply notched labium which is fused to the sternum separate this family from all other six-eyed, two-clawed spiders (Roth 1993).

Dysderids build tubular retreats in which to hide.

Biology/Taxonomic Status

The only species reported from Texas is *Dysdera crocota* C. L. Koch. This species is widespread and apparently is introduced from Europe.

Dysdera crocota C. L. Koch has the cephalothorax and legs reddish orange and the abdomen dirty white, with few hairs over the body (Figure 19). The chelicerae are long and project forward considerably, which typically is found only in the mygalomorph spiders and a few Salticidae. The six eyes are arranged in a transverse oval. Length of the female 11 to 15 mm; length of the male 9 to 10 mm.

No snare is built. These spiders construct flattened, oval retreats of tough silk from which they hunt prey. Eggs are laid in a very light, transparent eggsac. The spiderlings live with their mother for a while after they emerge. They eat sowbugs (Isopoda) and are often found with them. They have been suspected of mild cases of human envenomation.

They live under stones, sometimes under bark and in moss, for the most part seeking dark and humid surroundings.

There are few records from Texas. Also recorded from New England south to Georgia and west to California.

References

Kaston (1978). Forster and Platnick (1985).

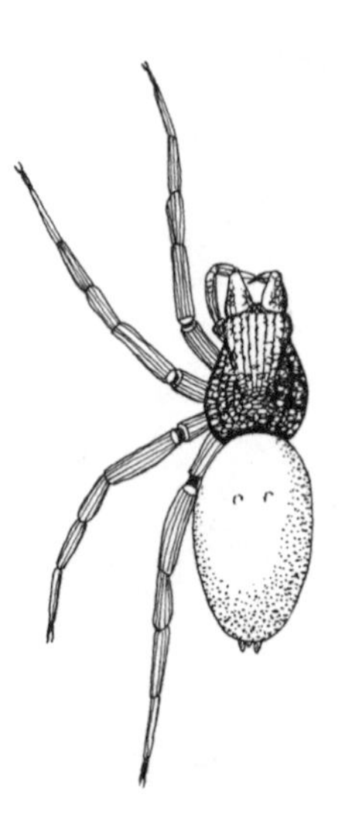

Figure 19. Dorsal view of *Dysdera crocota*. *(Redrawn from Kaston 1978.)*

13 MIMETIDAE—PIRATE SPIDERS

Key Family Characters

Pirate spiders are slow moving, with rather delicate legs. They generally appear yellow to whitish in color. The characteristic rows of strong, curved setae on the front margins of the lower segments of the first legs are usually adequate to distinguish this family from all others. The chelicerae are fairly long and slender and, at least for our species, are fused at the base.

Biology

Mimetids feed largely on other spiders. Pirate spiders bite their prey and feed by sucking on one leg of the prey. They commonly invade the webs of other spiders, but some sit with outstretched legs and wait for passing prey.

They have been taken from bushes and from under loose stones on the ground. The eggs are left in a stalked sac suspended from a twig or rock. These spiders are widespread but usually not very common.

Pirate spiders are not known to build snares of their own.

Taxonomic Status

This is a small family with only about a dozen species north of Mexico. All the Texas records are in the genera *Ero* and *Mimetus*. The genus *Ero* has a very high, rounded abdomen and a cephalothorax that is highest near the middle and slopes downward at the rear. *Ero* is widespread but less common than our species of *Mimetus*. *Ero* is smaller than most *Mimetus*, with a size of 2.3 to 3.4 mm. Breene et al. (1993b) have a key to species that occur in cotton fields. A revision of the genus *Mimetus* was last completed by Chamberlin (1923).

References

Levi et al. (1990). Breene et al. (1993b). Chamberlin (1923).

13A. *Mimetus hesperus* CHAMBERLIN

Description. *M. hesperus* is yellow with four thin black lines extending from the eyes and converging at the dorsal furrow on the cephalothorax. There are black spots under the first and second femorae, and two on the front of each chelicera. The abdomen has a series of irregular short lines in rows down the dorsum. There is also a pair of short longitudinal lines near the front of the abdomen.

Length of the female 4.0 to 6.3 mm; length of the male 3.5 to 4.5 mm.

Biology. *M. hesperus* has been reported preying upon black widow spiders, a small cobweb spider, *Theridion* sp., and *Dictyna* sp. (Agnew and Smith 1989).

Range. There are widespread records throughout most of Texas.

Habitat. Usually found on the underside of the leaf in the upper part of the cotton plant from June to August, it probably occurs in many other habitats.

Reference. Breene et al. (1993b).

13B. *Mimetus notius* Chamberlin

Description. The overall background color is yellow. The abdominal folium is a mass of curved, wavy, or zigzag black lines encompassing red markings. The cephalothorax has W-shaped black markings.

Length of the female 5 mm; length of the male 4 mm.

Biology. *M. notius* is largely found on the underside of leaves in the upper half of cotton plants.

Range. There are widespread records throughout Texas.

Habitat. Found on a wide variety of plants.

Taxonomic Status. A similar species, *M. puritanus* Chamberlin has a double Y-shaped mark on the carapace. The branched part of the Y is in the eye region. The border along the folium on the abdomen has a serrated black line and two comma-like pale or white marks between the "shoulders" on the abdomen. Length of the female 5.0 to 5.6 mm; length of the male 4.0 to 4.5 mm.

References. Breene et al. (1993b). Kaston (1978).

14 OECOBIIDAE—FLATMESH WEAVERS

Key Family Characters

Oecobiids are small spiders which have a large, hairy anal tubercle. The carapace and sternum are wider than long and quite rounded in appearance. They have three claws and a cribellum. The cribellum may be rudimentary and the calamistrum may be absent in males. The apical segment of the posterior spinnerets are upturned.

Biology

A few inhabit houses and may be found on window ledges, walls, and ceilings. Some feed on ants. These spiders rapidly circle their prey, wrapping it with silk to immobilize it.

They make small, flat webs about the size of postage stamps over crevices on walls or on leaves. Their tent-like webs with 5–7 anchor lines are often seen on inside corners of buildings. The webs are also placed over depressions on the undersides of rocks.

References

Roth (1993). Kaston (1978). Shear (1970). Coddington and Levi (1991).

14A. *Oecobius annulipes* LUCAS

Description. The carapace is pale yellow with scattered dark spots. There is also a marginal black line on the carapace. The abdomen is white or light brown with many dark spots, and the underside is pale. The calamistrum has a double row of bristles which extend about half the length of the fourth metatarsus.

Length of the female 2.5 mm; length of the male 2 mm.

Biology. This spider makes a small, flat web on window sills and over cracks on the walls of buildings. They may be very abundant at times.

Range. Recorded from central, west, and south Texas, New England south to Florida, and west to the Pacific.

Habitat. Often found in buildings.

Taxonomic Status. This species is very similar to O. *cellariorum* (Duges), which has similar size and habits. It is recorded from central and north central Texas.

Outdated and Unofficial Names. *Oecobius parietalis* (Hentz), O. *navus* Blackwall.

References. Kaston (1978). Levi et al. (1990). Platnick (1993). Lehtinen (1967).

15 HERSILIIDAE—LONGSPINNERET SPIDERS

Key Family Characters

These cryptically colored, long-legged spiders sit head down on tree trunks or stone walls. The third leg is by far the shortest, and the long, tapering spinnerets are very conspicuous. The posterior spinnerets are as long as the abdomen.

Biology

When an insect approaches the waiting spider, the spider jumps over the prey. The spider then spreads silk by rapidly running in circles with the spinnerets toward the prey. When the prey is completely wrapped, it is bitten and eaten.

These flattened, camouflaged spiders are active at night on tree trunks. They are amazingly fast.

No snare is built.

Taxonomic Status

This is primarily a tropical and subtropical family with only about 50 species. The only species in Texas is *Tama mexicana* (O. P.-Cambridge) which occurs only in the southernmost tip of Texas. Length is about 12 mm.

References

Levi et al. (1990). Comstock (1940).

6 Uloboridae—Hackled Orbweavers

Key Family Characters

Females have a well-developed pair of small humps at the highest point of the abdomen and a well-developed calamistrum on the fourth leg. The feathery protrusions on the distal section of the tibia of leg 1 of female uloborids are also useful in separating this group.

Biology

These spiders make geometric orb webs or sectors of orbs. In appearance, the spiders do not look much like the orbweavers in the family Araneidae. They are unusual among spiders because they have no poison glands. Some species in this family are social, but our species are not.

Webs of uloborids may be distinguished from most other orbweaving spiders by the horizontally oriented web. Some tetragnathids spin a horizontal web, but most of them, and most araneid webs, are nearly vertical. Webs are approximately 10 to 15 mm in diameter and are usually built at the middle of a plant. Another feature of the webs is the feathered or messy appearance of individual web strands, which the spider creates by combing the web with the calamistrum.

Taxonomic Status

Only a few species in this family occur in North America. Muma and Gertsch (1964) revised the family in North America. Subsequent work by Opell (1979) is also useful.

Reference

Breene et al. (1993b).

16A. *Hyptiotes cavatus* (Hentz), triangle weaver

Description. The abdomen of the female is broadly elliptical and has a double row of rounded bumps. The male has a narrower abdomen, and the bumps are less prominent. Males have the palpal organ very large and projecting. The body color resembles the brownish-gray ends of the branches among which it lives.

Length of the female 2.3 to 4 mm; length of the male 2 to 2.6 mm.

Biology. The web is easily recognized because of its triangular shape as a sector of about 45 degrees of an orb in a vertical plane. Connecting threads are hackled to aid in capture of prey. The line upon which the four radii converge is attached to a twig, and the spider usually stands, upside down, quite close to the twig, resembling a bud. The spider attaches itself to a twig, and holds the web taut. When prey falls on the web, the spider releases tension on the web which further helps to entangle the prey.

Range. East Texas. New England to Georgia and west to Texas and Missouri.

Habitat. Found in deciduous trees, underbrush, and pinewood habitats.

References. Kaston (1978). Levi et al. (1990). Gertsch (1979).

16B. *Uloborus glomosus* (Walckenaer)

FEATHERLEGGED ORBWEAVER

Description. The tibiae of the first leg of females bear a clump of hairs at their distal ends. The abdomen of the female is elevated with a pair of humps about one-third of its length from the notched anterior end. The carapace of the female is about one-and-one-third times as long as broad. The posterior row of eyes is so strongly recurved that a line along the front edges of the lateral eyes would not touch the medians. The general color is a drab grayish brown.

Length of the female 2.8 to 5 mm; length of the male 2.3 to 3.2 mm.

Biology. Members of this genus spin a complete orb, almost always on a horizontal plane, about 10 to 15 cm in diameter. The hackled silk is used rather than sticky threads to capture prey. The web is strengthened with a stabilimentum, or sheeted hub. The spider does not build a retreat and hangs below the web.

The eggsacs are elongate and light brown, about 6 mm in length with several papillae, and are suspended from or near the web. Each eggsac has about 50 eggs, and several occupy a single web.

In an east Texas cotton field, the prey of this species consisted largely of aphids that fell from leaves above the web (Nyffeler et al. 1989). These spiders were observed capturing predominantly adult flies and whiteflies in other habitats (Muma 1975).

They reach maturity in early summer and are usually found from July through September.

Range. Uloborids are more numerous in the eastern half of Texas. Found throughout the United States.

Habitat. Wooded areas.

References. Kaston (1978). Levi et al. (1990). Breene et al. (1993b).

17 NESTICIDAE—CAVE COBWEB SPIDERS

Key Family Characters

Nesticids are aerial web-spinners that are ecribellate, eight-eyed, three-clawed spiders with a comb of curved serrated bristles on tarsus 4. This set of characters is shared by the Theridiidae and the two families can easily be confused. These small spiders are usually light in color. The eyes are reduced in size or obsolete.

Biology

These sedentary spiders are usually found hanging upside down in small, irregular, tangled webs in soil crevices, leaf litter and other detritus, or caves. They may stray short distances from the webs. Eggsacs are carried on the spinnerets supported by the hind leg.

This family is a predominately northern one, and most of our species are found in the Edwards Plateau and karst areas of central Texas.

Taxonomic Status

Gertsch (1984) published a revision of this family.

References

Gertsch (1984).

17A. *Eidmannella pallida* (Emerton)

Description. This small spider has orange legs, an orange carapace, and a grayish abdomen.

Length of the female 2.2 to 4.0 mm; length of the male 2.2 to 2.8 mm.

Biology. Cave spiders typically build their loosely meshed webs in protected crevices. They hang upside down in their webs and construct eggsacs with as many as 96 eggs. They attach eggsacs to their spinnerets or keep closely by them in the web. The sacs are spherical, 4 mm in diameter, and thinly covered with whitish transparent silk.

Range. Widespread in Texas.

Habitat. Usually located in ground litter or near the soil. Sometimes found in cotton fields but also reported in caves.

Reference. Breene et al. (1993b).

Key Family Characters

The hind tarsi of theridiids have a row of six to ten slightly curved serrated bristles composing a comb on the ventral surface which is the most distinct feature of the family (Figure 20). However, this comb may be inconspicuous, difficult to see, or even absent in some specimens. The comb is also found in the Nesticidae. The comb can only be recognized with magnification. The spiders use the comb to throw viscid silk over prey when it lands in the webs.

Theridiids have eight eyes, three claws, and no cribellum. Most theridiids lack strong setae on the legs. Many have spherical abdomens, but some have the abdomen elongated or even pointed.

Biology

This is a large family with perhaps 200 species in the United States. They are some of our most common spiders. They seem to have poor vision. Males may pluck the threads of a female's web while courting.

Theridiids build tangled, irregular snares and suspend themselves in an inverted position while awaiting their prey. Webs have tunnel-like areas and passageways even though they are asymmetric. Viscid silk is flung over the prey using the hind legs.

Only after the prey is enswathed and relatively quiet does the spider approach close enough to bite it. Most commonly, the victim is dragged to that part of the web in which the spider usually resides. Theridiids sometimes spend a long time subduing prey that is many times larger than themselves.

Theridiids constitute an important part of our spider fauna. They are important predators in nearly every habitat.

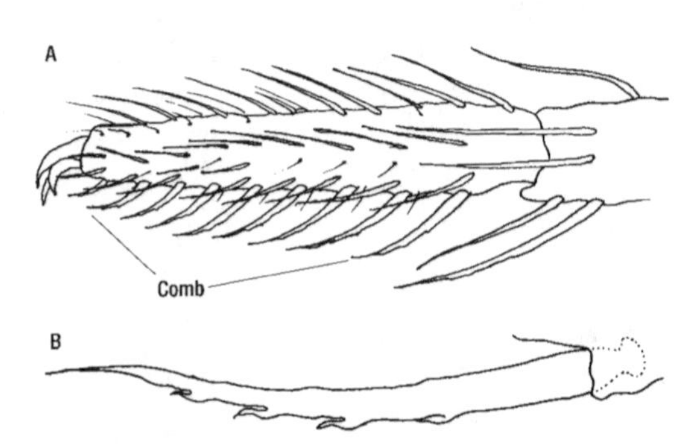

Figure 20. Hind tarsus of *Theridion* showing comb of serrated bristles (A) and a single bristle enlarged (B).

Taxonomic Status

With few exceptions, theridiid genera are difficult to distinguish. Many of the species have rounded abdomens, are dull colored, and have mottled or irregular patterns which may vary somewhat.

Revisions of the theridiid genera can be found in Exline and Levi (1962) and Levi (1954, 1955a, 1955b, 1956, 1957a, 1957b, 1959a, 1959b). Levi and Randolph (1975) present a key and checklist to the American Theridiidae.

Outdated and Unofficial Names

Cobweb spiders, combfooted spiders.

References

Kaston (1978). Levi et al. (1990). Breene et al. (1993b). Roth (1993).

Generic Summary for Theridiidae in Texas

Quick recognition characters	Genus	Number of species
Red hourglass or triangles on the underside	*Latrodectus*	4
Often with silver spots on the abdomen; pointed or elongate abdomen	*Argyrodes*	12
Pointed or angular abdomen shape	*Chrysso*	1
	Euryopis	6
	Spintharus	1
Other genera not easily recognized		
	Achaearanea	6
	Anelosimus	1
	Chrosiothes	2
	Coleosoma	2
	Crustulina	2
	Dipoena	4
	Enoplognatha	2
	Episinus	1
	Nesticoides	1
	Phoroncidia	1
	Steatoda	9
	Stemmops	1
	Theridion	22
	Theridula	1
	Thymoites	6
	Tidarren	2
	Wamba	1

18A. *Achaearanea tepidariorum* (C. L. Koch)
COMMON HOUSE SPIDER

Description. The carapace is yellowish brown, and the abdomen is dirty white to brown with indistinct gray chevrons on the posterior half. The legs of the male are orange; the legs of the female are yellow with dusky rings at the ends of the segments. Some individuals have a conspicuous spot in the center of the dorsum of the abdomen.

Length of the female 5 to 6 mm; length of the male 3.8 to 4.7 mm.

Biology. *A. tepidariorum* is probably the spider most often observed by humans in the United States. This extremely common spider is found most often in barns and houses where it makes its webs in the corners of rooms and in the angles of windows. The common house spider is not considered harmful to humans, but has roughly the same shape as the black widow spider.

Adults can be found at all seasons, and some individuals may live for a year or more after becoming mature. The eggsacs are familiar objects in the webs. They are brownish, ovoid or suborbicular, sometimes pear-shaped, 6 to 9 mm in diameter, with a tough, papery cover.

A. tepidariorum was observed trapping red imported fire ants in cotton fields. *Achaearanea* often capture various ants and beetles. The web is an irregular cobweb type. It usually stands in a more closely woven part of the web, but does not construct a "tent."

Range. Throughout the United States.

Habitat. Abandoned webs (cobwebs) inside human habitations are often made by this species. It has been collected outside under stones and boards, and on bridges and fences.

Taxonomic Status. *Achaearanea* has several other species in the genus which are worthy of mention.

A. porteri (Banks) has a yellow to brown carapace which is darker along the margins and in the ocular area. The abdomen is irregularly spotted with black and gray. Length of the female 2.2 to 4.9 mm; length of the male 1.6 to 2.8 mm.

A. globosa (Hentz) has the cephalothorax and legs orange. The dorsum of the abdomen from the midpoint to the spinnerets is whitish with black markings. Length of the female 1.0 to 2.2 mm; length of the male 1.1 to 1.7 mm.

Outdated and Unofficial Names. *Archaearanea tepidariorum,* house spider, domestic spider.

References. Kaston (1978). Levi et al. (1990). Platnick (1989, 1993). Breene et al. (1993b).

18B. *Anelosimus studiosus* (Hentz)

Description. The carapace and legs are orange-yellow with an indistinct median gray band extending forward from the dorsal groove, forking there with two branches toward the eyes. The abdomen is gray along the sides and has a dark median band bordered in white.

Length of the female 3.2 to 4.7 mm; length of the male 2.1 to 3.3 mm.

Biology. Spiders are typically solitary animals and highly cannibalistic. *A. studiosus* is one of the exceptions in that it forms colonies. Colonies can be found consistently in branches of live oak trees along the shores of Lake Somerville. The web is a platform sheet with irregular capture threads spun above it into which potential prey fly or fall. Muma (1975) found the carcasses of numerous adult midges in the webs of *A. studiosus* and concluded that midges probably form a large part of its diet in areas favorable to midge development.

An adult female typically initiates a colony. She begins a nest web alone and produces eggsacs containing approximately 30 to 50 eggs, which she will guard (Buskirk 1981). After spiderlings emerge, the mother feeds them with her regurgitated food. Later in their development, she will supply the spiderlings with the prey she has captured. As the spiderlings mature, they begin to assist the mother in securing prey. More information on the biology of *A. studiosus* is in Brach (1977).

Range. Eastern half of Texas. New England to Florida and west to California.

Habitat. These spiders typically occupy forested regions on tree limbs but also appear on low vegetation. Occasionally found in cotton in the eastern half of Texas from April to October.

Outdated and Unofficial Names. *Anelosimus textrix* (Walckenaer).

References. Kaston (1978). Breene et al. (1993b).

18C. *Argyrodes trigonum* (Hentz)

Description. *A. trigonum* is most commonly reddish brown or brownish yellow with a triangular abdomen, sometimes with metallic reflections. The abdomen is drawn out beyond the spinnerets, and though the back is usually straight, it has been said that the spider can turn down the tip. The ocular area of the female is raised and separated from the clypeus by a notch. The male has a long horn on either side of this notch and the abdomen is less angular than in the female.

Length of the female 3.7 to 4.2 mm; length of the male 2.4 to 3.3 mm.

Biology. Although *A. trigonum* can build their own webs, they are more commonly observed in the webs of other spiders. They occupy the periphery of webs of orbweaving spiders of the genus *Araneus.* They are considered kleptoparasitic, i.e., stealing prey caught on the web of the orbweaver or taking prey already captured and wrapped by the host spider. Occasionally, *A. trigonum* may feed on the web owner.

The eggsacs (6 mm long with 15 to 49 eggs) are distinctively urn-shaped and suspended from the web by a silk thread. The color of the eggsac changes from white when new to brown with age.

Range. Eastern Texas.

Habitat. This species is found upon the webs of orbweavers. They can be found from July to September.

Taxonomic Status. Members of the genus *Argyrodes* are called dewdrop spiders because of the silver coloration. Individuals of the silver-colored *A. elevatus* Taczanowski have been observed in the orbwebs of *Argiope aurantia* (Nyffeler et al. 1987b).

Outdated and Unofficial Names. *Conopistha rufa* (Walckenaer).

References. Breene et al. (1993b). Kaston (1978).

18D. *Latrodectus,* WIDOW SPIDERS

Biology. There are more than 25 *Latrodectus* species worldwide. Contrary to common belief, the female does not consume the male in most situations (Breene and Sweet 1985, Williams et al. 1986), except when held together in cages from which the male cannot escape. The Australian red-backed spider, *Latrodectus hasselti,* is the best known species in which the male of the species is typically killed by the female after mating (Forster 1992).

Latrodectus are commonly found among the spiders hunted by mud dauber wasps (Hymenoptera: Sphecidae), which capture and paralyze the spiders with their sting. The wasp lays an egg at the blind end of the mud cell, which is then provisioned with paralyzed spiders. The mud dauber egg hatches, and the young larva uses the spiders as its food source until it finally pupates and emerges from the cell (Dean et al. 1988).

Taxonomic Status. The three common species in Texas can be separated to some degree by the shape and location of the red markings on the abdomen. Experts generally agree on the separation of *L. hesperus* Chamberlin & Ivie and *L. mactans* (Fabricius). However, they do not always agree on the status of *L. variolus* Walckenaer. Examination of the genitalia is the proper way to identify species in the group.

The brown widow, *L. geometricus* C. L. Koch, is dark to light brown with a pattern. The hourglass is orange. Brown widows are sometimes called "gas station spiders" because of their habit of building webs in service stations. They are extending their range westward from Florida and Georgia, and have recently been reported in Texas.

The red widow, *L. bishopi* Kaston, is restricted to S. Florida.

Levi (1959a) published a revision of *Latrodectus*.

References. Levi (1959a). Kaston (1978). Breene et al. (1993b).

18E. *Latrodectus hesperus* Chamberlin & Ivie

WESTERN BLACK WIDOW

Description. The western black widow usually has the hourglass marking connected or complete with the anterior triangle larger and wider than the posterior triangle (Figures 21, 22, and 23).

Figure 21. Dorsal view of the body of a female *Latrodectus hesperus.* *(Redrawn from Kaston 1978.)*

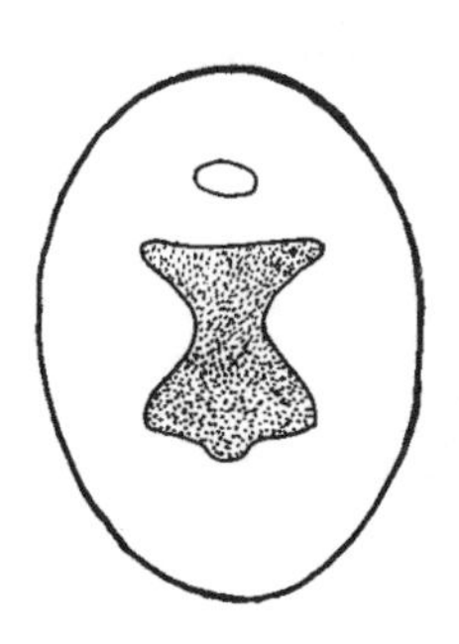

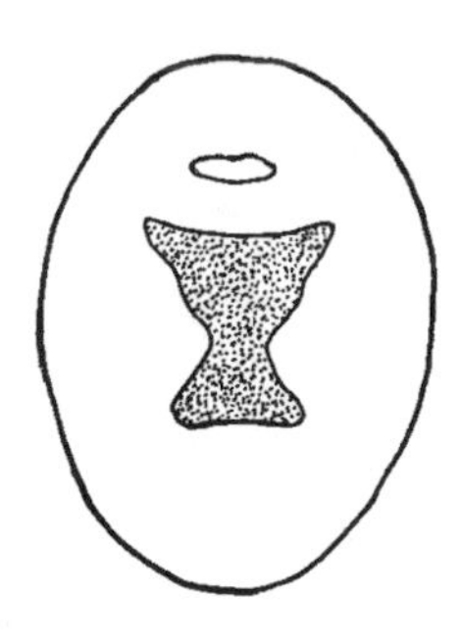

Figure 22. Ventral view of the abdomen of a female *Latrodectus hesperus* showing two types of hourglass markings. *(Redrawn from Kaston 1978.)*

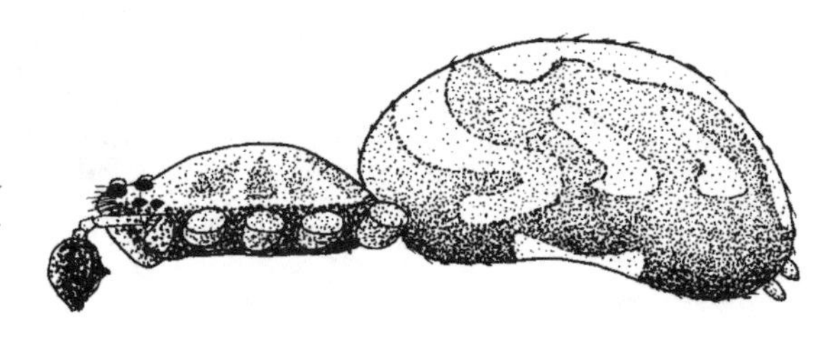

Figure 23. Side view of a male *Latrodectus hesperus.* *(Redrawn from Kaston 1978.)*

THERIDIIDAE—COBWEB WEAVERS

Biology. *L. hesperus* has been observed feeding on ants (MacKay 1982).

Range. This species largely displaces *L. mactans,* the southern black widow, in the western half of the state.

Habitat. This species can be found almost anywhere from ground level to high in trees and hedges.

Taxonomic Status. In southwestern Texas through the Lower Rio Grande Valley and adjoining parts of Mexico, specimens of *L. hesperus* have been found in which the adults retain their brilliant immature colors. Farther west, the coloration of the species appears to grade back to black.

References. Kaston (1978).

18F. *Latrodectus mactans* (FABRICIUS), SOUTHERN BLACK WIDOW

Description. The body of the female is shiny black with red spots. There is usually an hourglass-shaped mark on the venter, but this may be reduced to remnants. Usually there is a single red spot just behind the spinnerets, sometimes a row along the back, and sometimes all spots are missing entirely. There is much variation, with southern and western specimens being more strikingly marked than northern and eastern ones. The male has the abdomen narrower and with white lines along the sides. Immature females show a similar pattern. Young spiderlings are orange and white, and acquire more black in successive instars until in the adult stage they have little if any red remaining except the hourglass markings. The abdomen of immatures is gray with curved white stripes (Figures 24, 25, and 26).

Figure 24. Dorsal view of the body of a female *Latrodectus mactans. (Redrawn from Kaston 1978.)*

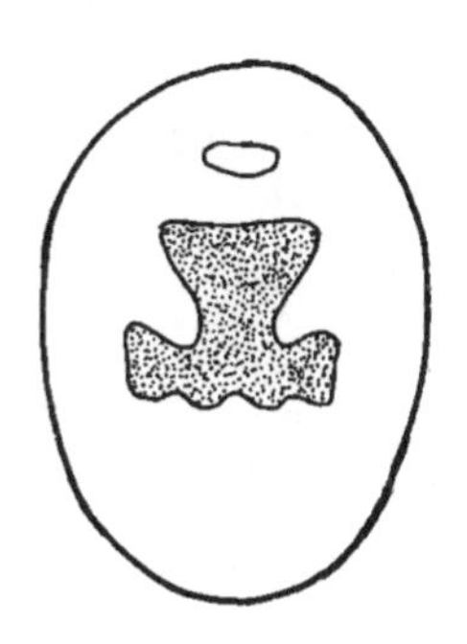

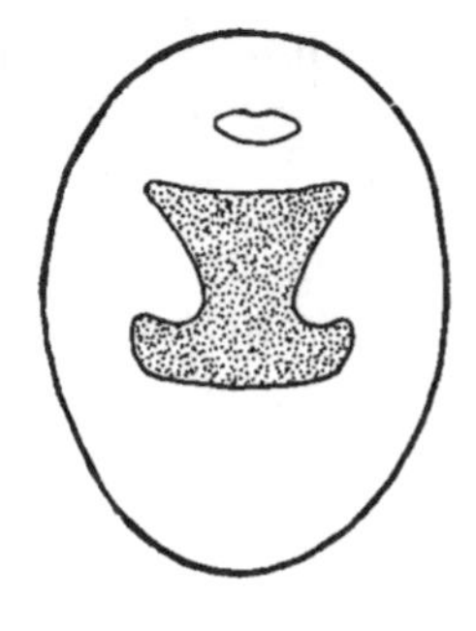

Figure 25. Ventral view of the abdomen of a female *Latrodectus mactans* showing two types of hourglass markings. *(Redrawn from Kaston 1978.)*

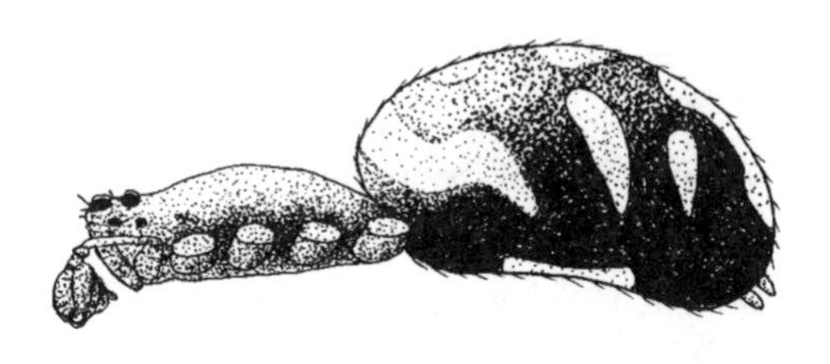

Figure 26. Side view of a male *Latrodectus mactans.* *(Redrawn from Kaston 1978.)*

Length of the female 8 to 10 mm; length of the male 3 to 6 mm, but sizes within different geographical populations can vary widely.

Biology. This species is perhaps the most notorious of all spiders in the United States. It is true that its venom is highly virulent, but the spider is quite timid. Even when disturbed in its web, it attempts to escape rather than attack.

Its web is an irregular mesh usually built in a dark spot sheltered from the weather. The web may also have a retreat, typically a 2 mm to 8 cm circular or semicircular silken tent. The spider spends most of the time in the retreat, venturing out onto the web for web maintenance or when attracted by prey vibrations. Webs are usually placed low to the ground.

During the course of a summer, a female may lay several eggsacs. The eggsacs are white to tan or gray, pear-shaped to almost globular, of tough, papery texture, and about 8 to 12 mm in diameter. According to Deevey (1949) and Williams et al. (1986a), eggsacs contain from 25 to 250 or more eggs per sac, although others have reported as many as 400 eggs. Eggsacs are suspended in the web where the female stands guard nearby.

The second instar spiderlings typically emerge about 4 weeks after eggsac production. Newly emerged spiderlings are not cannibalistic until 10 days to 2 weeks after emergence, whereupon they may suddenly become highly cannibalistic. Males generally require fewer molts to mature than females.

Red imported fire ants have been reported as their main food in cotton fields of East Texas (Nyffeler et al. 1988b). Boll weevils, grasshoppers, June beetles, and scorpions are also included on the large list of prey known for *L. mactans* (Whitcomb et al.1963).

Range. The species is known from every state (except Maine where it undoubtedly also occurs) and several Canadian provinces. It is uncommon in the north but quite abundant in the south and west.

Habitat. The webs are commonly found in spaces under stones or logs, in holes in dirt embankments, and in barns, houses, outhouses, cotton fields, trash, and dumps. Adults of both sexes have been found throughout the year in buildings. They can be very common at some locations and times.

References. Kaston (1978). Levi et al. (1990). Platnick (1993). Breene et al. (1993b).

18G. *Latrodectus variolus* Walckenaer, northern black widow

Description. The northern black widow usually has the hourglass divided, typically with red spots on the dorsum and white lines on the sides (Figures 27, 28, and 29).

Biology. The eggsac is brown and paper-like.

Range. This species occurs throughout much of the eastern half of the United States.

Habitat. These spiders usually are found in bushes or trees. They also occur in undisturbed woods, in stumps, or in stone walls.

References. Kaston (1978).

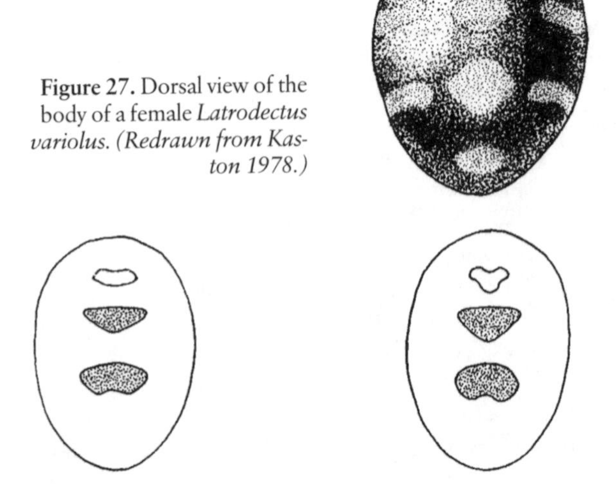

Figure 27. Dorsal view of the body of a female *Latrodectus variolus. (Redrawn from Kaston 1978.)*

Figure 28. Ventral view of the abdomen of a female *Latrodectus variolus* showing two types of hourglass markings. *(Redrawn from Kaston 1978.)*

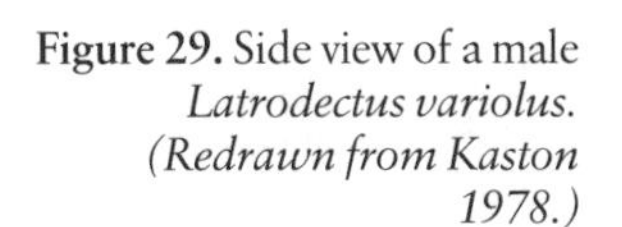
Figure 29. Side view of a male *Latrodectus variolus.* (*Redrawn from Kaston 1978.*)

18H. *Steatoda americana* (Emerton)

TWOSPOTTED COBWEB SPIDER

Description. The cephalothorax and legs are dark chestnut brown with the sternum somewhat darker than the carapace. The abdomen is dark purplish brown with a pair of white spots across the middle.

Length of the female 3.5 to 4.7 mm; length of the male 3.2 to 4.4 mm.

Biology. In buildings, *Steatoda* individuals have been observed killing detrimental insects including house flies, roaming larvae and adult mealworm beetles, and adults of various meal-infesting moths.

Most *Steatoda* species have a light band across the front of the abdominal dorsum. Males produce sounds during sexual and aggressive displays by scraping the carapace and abdomen together (Lee et al.1986, Nyffeler et al. 1986b).

Range. Reported from central and east Texas. Eastern United States and adjacent Canada to the Pacific Northwest.

Habitat. Found under stones and logs, and under loose bark and debris near the ground.

Outdated and Unofficial Names. *Asagena.*

References. Kaston (1978). Brignoli (1983). Platnick (1993). Breene et al. (1993b).

18I. *Steatoda triangulosa* (Walckenaer)

Description. The cephalothorax is brownish orange, with yellow legs grading darker toward each distal end of the segment. The abdomen is yellow with brown to purplish markings.

Length of the female 3.6 to 5.9 mm; length of the male 3.5 to 4.7 mm.

Biology. The eggsacs (5 mm in diameter) are made of loosely woven white silk, making the individual eggs (about 30) visible. This spider was reported to feed on the red imported fire ant (MacKay and Vinson 1989).

Range. This species can be found in the eastern half of Texas. Reported throughout the United States.

Habitat. *S. triangulosa* has been found under stones, bridges, culverts, and in buildings.

Outdated and Unofficial Names. *Steatoda teutana.*

References. Breene et al. (1993b). Kaston (1978).

18J. *Theridion*

Description. Most *Theridion* species have small bodies and make tiny webs. They hang upside down in the irregular webs. They can be found in nearly any habitat. The genus *Theridion* contains many species, with 23 species recorded in Texas. Identification is difficult, but a few of the more common species are mentioned below.

T. australe Banks has the carapace yellow to orange, except the ocular area which is blackened. The abdomen is orange-white with two black spots on the dorsum above the spinnerets. Length of the female 2.0 to 3.0 mm; length of the male 1.9 to 2.3 mm. It is the most commonly observed theridiid species in cotton ecosystems throughout Texas from May to September. *T. australe* prefers the upper parts of plants like cotton. It has been established as a predator of the cotton fleahopper on cotton (Breene et al. 1989a) and also feeds upon the red imported fire ant (Nyffeler et al. 1988a). Ants may be a primary food source, at least for immatures.

Theridion hidalgo Levi has a yellow-white carapace with a dark dusky or red band. The abdomen has a median scalloped white band on a gray spotted background. Length of the female 1.5 to 2.0 mm; length of the male 1.4 to 1.7 mm. It is found in the eastern half of Texas. It has been recorded from cotton fields.

Theridion murarium Emerton has the cephalothorax with black marginal stripes and a black median band running longitudinally on a background of grayish yellow. The abdomen has a lighter, wavy longitudinal band surrounded with darker regions. Length of the female 2.8 to 4.3 mm; length of the male 2.1 to 3.2 mm. The web is made in trees, in bushes and grass, and also under stones near the ground. The eggsacs are white to tan, spherical, about 3 to 4 mm in diameter, and at first are held by the hind legs then attached in the web and guarded. Eggsacs contain about 30 eggs.

T. murarium is widespread in Texas. Reported throughout the United States and southern Canada. The species is not common on cotton, where it has been found from June to August.

References. Kaston (1978). Levi et al. (1990). Platnick (1993). Breene et al. (1993b).

18K. *Tidarren haemorrhoidale* (Bertkau)

Description. Males of *Tidarren* are easily distinguished by the presence of a large single left pedipalp. The color is variable, from brownish tan to grayish black. The legs are spotted and banded. The female's abdomen is dirty white with brown or black markings and has a white vertical stripe on the posterior.

Length of the female 2.4 to 3.7 mm; length of the male 0.9 to 1.4 mm.

Biology. Dean et al. (1982) noted that the species was generally seen on the lower half of the cotton plant late in the season. The female builds a retreat, typically consisting of a curled leaf, often in the upper sections of the web, which she also uses for concealing her eggsacs. Fire ants may displace *Tidarren* because they are seldom recorded from cotton fields with fire ants.

Levi et al. (1990) stated that the male amputates one of his own disproportionally large pedipalps before his last molt.

Range. The eastern half of Texas.

Habitat. Usually found in the lower parts of plants like cotton.

Taxonomic Status. A similar species, *T. sisyphoides* (Walckenaer), is somewhat larger and occurs across the southern states.

References. Breene et al. (1993b).

18 LINYPHIIDAE—SHEETWEB AND DWARF WEAVERS

Key Family Characters

Linyphiids have eight eyes, three claws, and lack a cribellum. These spiders are small, with most species less than 3 mm. They lack the comb of theridiids but may be confused with them.

Two major subfamilies are recognized: Micryphantinae or dwarf spiders, and the Linyphiinae or sheetweb weavers. Micryphantinae are mostly under 2 mm long and make small webs. The males of dwarf spiders may have unusual turrets, bulges, or depressions in the head region. The abdomen is usually spherical and in some species is covered with a hard, shiny plate. Linyphiinae are generally larger than Micryphantinae. They usually have a patterned abdomen that is longer than wide.

Biology

These small spiders are among the most abundant in woodlands and grasslands but are often overlooked because of their small size. Moreover, this large family is probably less well-known than any other. Their prey is generally very small insects including springtails, flies, and leafhoppers.

Most species construct a snare of some sort, usually with a platform or dome, with an additional irregular portion. No retreat, molting, or eggsac webs are built. The spider takes up a position on the inner or under surface of the snare. Some fragile webs are placed over small depressions on the ground or in bushes. These spiders generally prefer shady areas.

Micryphantinae are abundant in leaf litter; others can be found under stones or collected with a sweep net. As many as 11,000 spiders per acre have been recorded in the United States with about two-thirds of them dwarf spiders.

Linyphiinae males and females may hang upside down under the same web and run rapidly when disturbed. The webs, which have few sticky threads, are found between branches of trees or bushes and in high grass. If an insect gets tangled in the web, the spider bites it from below the web. The web also protects the spider from predators. Only a few of the many species are widespread.

Taxonomic Status

The linyphiids may be difficult to separate from some of the theridiids.

Outdated and Unofficial Names

Line-weaving spiders, dwarf spiders.

References

Kaston (1978). Levi et al. (1990). Nyffeler et al. (1994). Breene et al. (1993b).

19A. *Frontinella communis* Hentz, bowl and doily weaver

Description. The carapace is evenly brown. The abdomen is about as high above the spinnerets as in front, and is marked above with a pattern.

Length of the female 3 to 4 mm; length of the male 3 to 3.3 mm.

Biology. The distinctive shape of the bowl and doily web gives the spider its common name.

Range. Found mostly in the eastern half of Texas. Reported throughout the United States and Canada.

Habitat. Found in pine woods, bushes, and tall grass.

Outdated and Unofficial Names. *Linyphia, Frontinella pyramitela* (Walckenaer).

References. Kaston (1978). Levi et al. (1990). Breene et al. (1993b).

19B. *Neriene radiata* (Walckenaer), filmy dome spider

Description. The carapace is dark brown except for the lateral margins which are white. The abdomen is widest and highest behind, with a characteristic pattern of brown markings on a yellowish white background.

Length of the female 4 to 5 mm; length of the male 3.5 to 4 mm.

Range. Widespread in Texas. New England south to Florida and west to California.

Habitat. The characteristic snares of this very common species are seen in wooded areas, in underbrush, and about rockpiles and stone walls.

Outdated and Unofficial Names. *Linyphia marginata* C. L. Koch, *Prolinyphia marginata* (C. L. Koch).

References. Kaston (1978). Levi et al. (1990).

19C. *Tennesseellum formica* (Emerton)

Description. The cephalothorax and legs are orange-yellow. The abdomen is white with a gray band encircling it near the front and another at the rear. In some specimens, the gray is quite indistinct on the dorsum though usually visible on the sides and venter. In others, the abdomen is almost all gray. In both sexes, a constriction or depression, more pronounced in males, is present at about the middle of the dorsum.

Length of the female 1.8 to 2.5 mm; length of the male 1.8 to 2.4 mm.

Range. This species is found in the eastern half of Texas from May through September. Reported from New England south to Georgia and west to California.

Habitat. Found among dead leaves on the forest floor.

References. Kaston (1978). Platnick (1993). Breene et al. (1993b).

.o TETRAGNATHIDAE—LONGJAWED ORBWEAVERS

Key Family Characters

Tetragnathids are usually large spiders with eight eyes, three claws, and no cribellum. They may be confused with araneids but usually have the the lateral eyes separated from the median eyes by about the width of the eye. The chelicerae are enlarged, especially in males of some genera. The first two legs, including the femora, are extended forward when at rest in most genera (Roth 1993). Females lack an epigynum in some genera.

Biology

Much of the literature states that the preferred habitat is near water. However, some species are nearly ubiquitous.

Most tetragnathids make orbwebs usually closer to horizontal than vertical. The webs usually have 12–20 radii and widely spaced spirals. As far as is known, the webs are complete orbs; there are relatively few radii and few spirals. The spider often hangs in the center of the web.

Taxonomic Status

The Tetragnathidae have been removed from, and then rejoined to, the orbweaver family Araneidae more than once in the past. There are about two dozen species north of Mexico.

References

Levi et al. (1990). Kaston (1978). Breene et al. (1993b). Revisions were completed by Levi (1980, 1981).

Generic Summary for Tetragnathidae in Texas

Quick recognition characters	Genus	Number of species
Large jaws; usually large spider	*Tetragnatha*	10
Shorter, thick jaws	*Pachygnatha*	1
Large, long-legged spider with bushy sections on the legs	*Nephila*	1
Other genera not easily recognized		
	Azilia	1
	Glenognatha	1
	Leucauge	1
	Meta	1

20A. *Leucauge venusta* (WALCKENAER), ORCHARD ORBWEAVER

Description. *L. venusta* has the carapace yellowish gray or green with darker stripes on the sides. The sternum is green. *Leucauge* have long, feathery trichobothira, proximally and on the anterior surface of the femur of the fourth leg (Figure 30). The labium is wider than or as wide as long. The legs are fairly long and the first leg longest. The abdomen is longer than wide, with the sides subparallel with silver coloration broken in several longitudinal lines. Males are only slightly smaller than females.

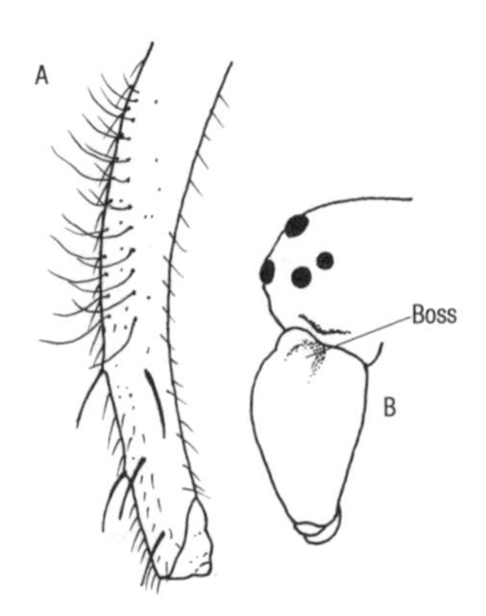

Figure 30. Hind femur (A) and chelicera (B) with rudimentary boss of *Leucauge. (Redrawn from Kaston 1978.)*

The abdomen is elliptical, iridescent above, and with dark lines (Figure 31). The iridescent color is described as gold or silver. The sides are yellow with red spots near the posterior end, and there is a red spot in the middle of the venter. The posterior of the abdomen is black, venter with two silver triangles with fused pigmented spots and additional scattered silver pigment on the outside of the triangles.

Length of the female 5.5 to 7.5 mm; length of the male 3.5 to 4 mm.

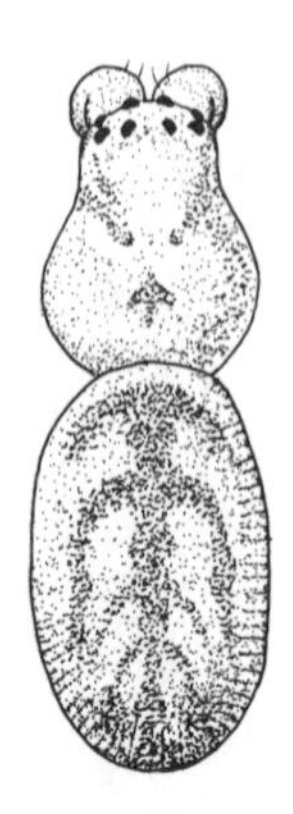

Figure 31. Dorsal view of the body of *Leucauge venusta. (Redrawn from Kaston 1978.)*

Biology. Webs are horizontal to nearly vertical, with an irregular barrier web below. The hub is open. There are approximately 30 radii or more, a wide free zone, and over 60 spirals. The spider sits in the web with the tip of the abdomen below the open hub.

Eggsacs are loose and fluffy, made of orange-white silk. They are 8–9 mm in diameter and hold several hundred eggs.

Range. Found in the eastern half of Texas. Reported from New England south to Florida and west to Texas and Nebraska.

Habitat. Common in wooded areas, low bushes, trees, and cotton fields.

References. Platnick (1993). Levi et al. (1990). Breene et al. (1993b). Levi (1980).

20B. *Nephila clavipes* (LINNAEUS), GOLDEN SILK ORBWEAVER

Description. This species is readily recognized by the conspicuous tufts of hairs on the femora and tibiae of legs 1, 2, and 4. The carapace is dark brown, the abdomen olive green with pairs of yellow and white spots. The labium is longer than wide. There is also a light line across the anterior end of the abdomen.

Length of the female 22 mm; length of the male 5 to 8 mm.

Biology. This species differs in some respects from others in the family. The webs are large and the structure is distinct. The radii are pulled out of their direct line to give a notched appearance, and the viscid spiral is yellowish rather than white in color (Figure 32).

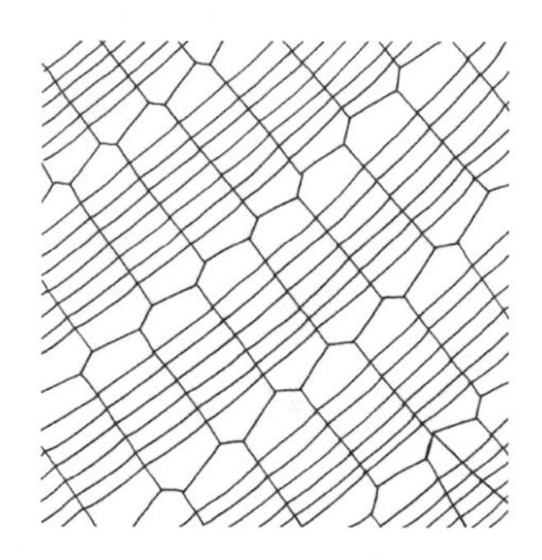

Figure 32. Close-up of a portion of the web of *Nephila*. *(Redrawn from Kaston 1978.)*

Range. There are only a few records from Texas; this species is probably more common along the Upper Gulf Coast than other parts of the state. Reported from Florida west to California.

Habitat. In shaded woods.

References. Kaston (1978). Platnick (1989, 1993). Levi (1980).

20C. *Tetragnatha*, LONGJAWED ORBWEAVERS

Description. The two eye rows may be parallel, they may diverge, or they may converge, but the lateral eyes are never contiguous. The chelicerae are long and well developed, especially in males, and the margins of the fang furrow have numerous teeth. In males, in addition to these marginal teeth, there is a strong projecting spur near the base of the fang. The abdomen is quite long and, in females, often swollen at the base.

Males cannot be separated to species reliably by the shape of the chelicerae.

Biology. At rest, these spiders may cling lengthwise to a twig or grass blade. They extend their long legs forward and back while clinging with their shorter third legs. They are common in meadows near water in North America.

Many species of *Tetragnatha* build their webs, which are often horizontal, in meadows and in bushes and long grass near water. The snare has few radii and an open hub on which the spider sits.

Taxonomic Status. There are ten species recorded in Texas, and they are difficult to distinguish. The long jaws and elongate bodies are distinctive for these large spiders.

References. Kaston (1978). Levi et al. (1990). Levi (1981).

20D. *Tetragnatha laboriosa* HENTZ
SILVER LONGJAWED ORBWEAVER

Description. The legs and carapace of *T. laboriosa* are yellowish, and the abdomen is elongate and silvery. The chelicerae are short and practically vertical, being one-half to two-thirds the length of the carapace in males and shorter in females. The lateral eyes of each side are as far apart as the medians.

Length of the female 5.2 to 9 mm; length of the male 3.8 to 7.4 mm.

Biology. This spider builds a new web every night and is most active at sundown. It only builds webs when conditions are warm and not windy.

At times, *T. laboriosa* can be the most abundant spider species on cotton, especially in western states. It is usually found from May to September. Its reproductive capabilities and ability to disperse by ballooning are remarkable. This spider often builds its web on the upper half of the cotton plant (Dean et al. 1982) or even between rows. It feeds on small insects, especially flies, leafhoppers, and aphids (Whitcomb et al. 1963, Nyffeler et al. 1989). The species was found to be a predator of immature cotton fleahoppers on woolly croton (Breene et al. 1988a), but not on cotton. It seems to reject beetles and may cut them from its web.

Forty to 76 eggs are deposited in the eggsacs. More information on *T. laboriosa* can be found in LeSar and Unzicker (1978) and Culin and Yeargan (1982).

Range. The species is ubiquitous in the United States and Canada to Alaska (Kaston 1978).

Habitat. This spider is one of the most common spiders in field crops in the United States. *T. versicolor* Walckenaer may also be found far from water.

References. Breene et al. (1993b). Platnick (1993). Sterling (1982). Levi (1981).

21 Araneidae—Orbweavers

Key Family Characters

Araneids have eight eyes, three claws, and no cribellum. They have a colulus, and often have spinose legs. They lack the serrated comb of theridiids.

They differ from most tetragnathids in that the lateral eyes often are separated from the median eyes by 2–3 times the eye diameter, with *Cyclosa* and *Mecynogea* being exceptions. The chelicerae are not enlarged and are similar in both sexes. The epigynum may have a scape or at least show three-dimensional sculpturing. All of these spiders make orb webs except *Kaira* and the bolas spiders, *Agathostichus* and *Mastophora*.

The first two femora are directed obliquely forward in the resting position, with the femur-patella joint flexed. The abdomen usually lacks silver or white spots, but strong spines and humps may be present. The males of some species are small and are often found in the webs of females (Roth 1993).

Biology

Almost all of our species of araneids spin snares in the form of orbs, which is what most people think of as spiderwebs. Most orbweavers that remain in the web hang upside down in the center. Some build a retreat away from the snare, but others remain at the center of the orb.

In many species, there is considerable sexual dimorphism, with the males often much smaller than the females. Males may have special clasping spines or spurs on the legs, and some differ from females in the shape of the cephalothorax as well as the abdomen. Males are not seen as frequently as females because of their smaller size and shorter life spans.

Orbweavers have poor vision. They locate prey by feeling vibrations and tension on the web. They extrude silk and use their hind legs to wrap the prey. Their prey is bitten before being carried to the center of the web or to the spider's retreat in a corner of the web. Nonedible portions of the prey will be cut from the web and discarded.

Many species reach maturity in the late summer or fall, when eggsacs are laid. Eggs may hatch in the fall or spring.

Orbweavers have been used in behavioral research to test the effects of gravity, disturbances, and drugs on web building.

Moths and butterflies are not easily captured in orb webs. The loose scales on their wings may allow them to escape from the sticky webs, but the legs and body may still get caught (Eisner et al. 1964).

An orb web consists of foundation lines that form the outer framework of the web. The center of the web is called the hub. Radii connect the hub with the foundation lines. Sticky or viscid silk is spun across the radii in spirals. Orb webs can have open sectors or various attachments which are somewhat characteristic of the species. Some orbweavers produce a stabilimentum, which is a thickened region of the web usually near the hub that may be in a straight line or zigzag pattern. There are several proposed theories for the function of the stabilimentum including: to strengthen the web, as camouflage, as a molting platform, as a shield against radiation from the sun, or to repel birds which might damage the web accidentally. A more recent study proposes that the stabilimentum may reflect the ultraviolet frequency of flowers, thus acting to attract insect prey (Craig and Bernard 1990).

To construct a web, the spider first strings the upper foundation line or bridge line between two objects. Spiders use air currents to carry the thread from one object to another to lay the bridge line. After that, the spider uses the bridge line to lay the rest of the foundation lines. Radii are spun after the foundation is complete. The spider pulls on the web while building and thus senses how taut the lines are before they are fastened. Some orbweavers build a new web every night, while others use the same web for weeks.

Taxonomic Status

Further information on this family may be found in various revisions: Berman and Levi (1971), Edwards (1986), Levi (1968, 1970, 1971a, 1973, 1974, 1975, 1976, 1977, 1978, and 1980).

Outdated and Unofficial Names

Argiopidae (which included Theridiosomatidae and Tetragnathidae).

References

Kaston (1978). Levi et al. (1990). Breene et al. (1993b).

Generic Summary for Araneidae in Texas

Quick recognition characters	Genus	Number of species
Very large, long-legged, boldly colored	*Argiope*	4
Hard back; abdomen flattened; short, stout spines	*Gasteracantha*	1
Spiny but more fleshy abdomen	*Acanthepeira*	2
	Micrathena	3
Large, stout body; heavy spines on legs	*Neoscona*	6
Females with long epigynal scape	*Eriophora*	2
Silver-striped abdomen	*Mecynogea*	1
Round, heavy body; common, large, and smaller forms	*Araneus*	16
Dull colored to black; conspicuous caudal projection on abdomen; may have white on part of abdomen	*Cyclosa*	5
Feathery hairs on leg 4	*Mangora*	6

Continued on next page

Generic Summary for Araneidae in Texas *(continued)*

Other genera not easily recognized	Genus	Number of species
	Acacesia	1
	Agathostichus	1
	Araniella	1
	Colphepeira	1
	Eustala	8
	Gea	1
	Hypsosinga	2
	Kaira	3
	Larinia	1
	Larinioides	3
	Mastophora	3
	Metazygia	2
	Metepeira	5
	Ocrepeira	4
	Scoloderus	1
	Verrucosa	1
	Wagneriana	1

21A. *Acacesia hamata* (HENTZ)

Description. The carapace and legs are greenish gray to brown. The abdomen has a ground color of gray to green with black and white lines.

Length of the female 4.7 to 9.1 mm; length of the male 3.6 to 5 mm.

Biology. Like some other orbweavers, this species makes the web at sundown and removes it before sunrise.

Range. Southern half of Texas extending into east and central Texas. Reported from New England south to Florida and west to Illinois and Texas.

Habitat. In bushes and shaded wooded areas.

References. Kaston (1978). Platnick (1993). Breene et al. (1993b).

21B. *Acanthepeira stellata* (WALCKENAER)

STARBELLIED ORBWEAVER

Description. The carapace is brown; the legs are yellow and annulated with brown. The abdomen is brown, lighter on the posterior two-thirds and with dark spots. A median tubercle on the abdomen has a white spot and overhangs the carapace.

The male has two apical dorsal spikes on the palpal patella. Its abdomen is highly sclerotized (hardened), with 12 cones radiating from lateral areas. Its body is brown overall, and the legs are yellow with brown rings.

Length of the female 7 to 15 mm; length of the male 5 to 8 mm.

Biology. This species can at times be one of the most abundant orbweavers in cotton fields. Individuals may be found in the web at midday but will usually occupy a retreat at the edge of the web. The web is 15 to 25 cm in diameter and is typically built on the upper half of the cotton plant, often across rows.

Range. This species is found in the eastern two-thirds of Texas from May through September. Reported in New England and adjacent Canada south to Florida and west to Kansas and Arizona.

Habitat. In tall grass and low bushes.

References. Breene et al. (1993b). Kaston (1978). Sterling (1982).

21C. *Araneus*

Description. This large genus contains both large, heavy-bodied forms and smaller ones. There are "angulate" forms with well-developed shoulder humps and "round-shouldered" forms that lack the humps.

Males are short-lived and can only be found for brief periods once they are mature. Males of some species are unknown. They can use each palp only once for mating. Sphecid wasps often provision their nests with these spiders, and specimens can be taken by raiding the nests.

Araneus are some of our most common orbweavers. This genus is difficult to distinguish from *Neoscona,* which superficially has heavier spines on the legs. The genera *Araniella* and *Larinioides* are also similar.

The "large" species (over 8 mm for females), or *diadematus* groups, is represented in Texas by the following species: *bicentarius* (McCook), *cavaticus* (Keyserling), *illaudatus* (Gertsch & Mulaik), *marmoreus* (Clerck), and *nordmanni* (Thorell). The remaining species, including *A. pratensis,* are all "small" species (under 8 mm for females).

References. Levi (1971, 1973).

21D. *Araneus detrimentosus* (O. P.-Cambridge)

Description. The carapace and sternum are light brown. The carapace has some white setae. The legs are dark with bands. Dorsum of the abdomen is bright green underlaid by white. There is a prominent white band on each side of the anterior abdomen and a reddish brown band under the white band. The sides of the abdomen are light brown. The venter is light brown with four white, indistinct patches. The abdomen is wider than long. While females are usually bright green, sometimes the abdomen is white with brown marks. Males are colored like females.

Length of the female 5.3 mm; length of the male 3.8 mm.

Biology. These spiders can be collected from March to October. They spin an orb web about the size of a human palm, with a concave shelter.

Range. East, central, and south Texas. Reported from coastal states from Florida to California.

Habitat. On mesquite, mostly defoliated bushes, on juniper and elm, under reeds, along lagoons, and on hillsides.

References. Levi (1973).

21E. *Araneus marmoreus* CLERCK, MARBLED ORBWEAVER

Description. The cephalothorax is yellow with darker lines along the sides and in the middle. The abdomen is yellow-orange with brown to purple markings in a definite pattern. In the male, coxa 1 has a spur along its posterior edge and femur 2 has a groove along its anterior edge. Coxa 2 also has a spur and tibia 2 is much thickened and curved.

Length of the female 9 to 18 mm; length of the male 5.9 to 9 mm.

Biology. A signal line connects the hub with a retreat, made by fastening several leaves together somewhere above the snare.

Range. Most records are from east Texas. Reported throughout the United States.

Habitat. The webs are built in wooded areas between trees and shrubs or in meadows.

References. Kaston (1978). Levi et al. (1990). Platnick (1993).

21F. *Araneus pratensis* (EMERTON)

Description. The carapace of the spider is yellow in the middle and fades to light brown on the sides. The abdomen is light colored with two dark longitudinal stripes on the top. The stripes are brown in the front and darker toward the rear.

Length of the female 3.8 to 5.6 mm; length of the male 3.2 to 4 mm.

Range. There are scattered records from across Texas, but most are from eastern, southern, and central Texas. Reported in the eastern United States to Iowa and Texas.

Habitat. Found in tall grass and low trees.

Reference. Kaston (1978).

21G. *Araniella displicata* (Hentz), sixspotted orbweaver

Description. The color is variable and may appear greenish, reddish, brownish, or yellow. The carapace is yellow to brown without markings. It is smooth with almost no setae and no thoracic depression. Eyes are on black spots. Legs are yellow to brown. The abdomen is white, yellow, or pink, with three pairs of black spots on the posterior half. The abdominal pattern of lines and spots becomes more distinct in later instars. The abdomen is oval and has no folium pattern.

Males have a hook on the distal margin of the first coxa, and the second femur has a matching depression. Legs of males are longer than those of females.

Length of the female 4.8 to 7.2 mm; length of the male 4 to 5 mm.

Biology. This species was not found on Texas cotton until the late 1980s, when it was found to be a predator of cotton fleahoppers (Breene et al. 1989b). Wheeler (1973) and McCaffrey and Horsburgh (1980) listed prey species of *A. displicata* in habitats other than cotton.

Range. East, central, and south Texas. Reported from New England south to North Carolina and west to Minnesota; also the Rockies west to the Pacific Coast states, and all of Canada to Alaska. The species is more common in northern states.

Habitat. In tall grass and bushes.

Taxonomic Status. There are no good superficial characters to separate *Araniella* from the small *Araneus*. However, the three pairs of black spots on the abdomen are quite distinctive for this species.

References. Kaston (1978). Levi et al. (1990). Platnick (1993). Breene et al. (1993b). Levi (1974).

21H. *Argiope aurantia* Lucas, yellow garden spider

Description. The cephalothorax is covered with silvery hairs. The abdomen is slightly pointed behind and notched in front to form a hump on each side, and is marked in black and yellow (occasionally orange). The front legs are sometimes black, and sometimes have a short band of orange on the femur. Others have the femora reddish or yellow and the rest black.

Length of the female 19 to 28 mm; length of the male 5 to 8 mm.

Biology. Because of its large size, bright colors, and habit of sitting at the center of its web built in open, sunny places, this spider is one of the most noticed spiders in Texas. It may be found along roadsides, in cotton fields, and around homes. *Argiope* can kill prey at least twice its length, and feeds on several types of grasshoppers (Nyffeler et al. 1994). One or more males may be observed on the upper part of the penultimate female's web in August or later, waiting for her to molt into adulthood when courtship and mating can take place. The eggsacs, containing 400 to 1,000 or more eggs, are pear-shaped, brownish, papery, and pointed at the apex.

Some species of *Argiope* construct stabilimenta vertically on either side of the center hub of their webs. *Argiope* spiders may feed on honeybees, grasshoppers, and lepidopterans. They often exhibit a specialized predatory behavior toward adult lepidopterans (Robinson 1969). Predation behavior on cotton was studied by Nyffeler et al. (1987b) and Harwood (1974) in other habitats.

Range. The species is found in the eastern two-thirds of Texas and has been collected from June to August. Reported throughout the United States, but apparently not common in the Rockies and the Great Basin areas.

Habitat. The species can be found in old fields and minimally disturbed grasslands, along fence rows, and around homes. It builds webs in gardens around houses and in tall grass.

Outdated and Unofficial Names. Black and yellow garden spider, writing spider.

References. Kaston (1978). Levi et al. (1990). Breene et al. (1993b).

211. *Argiope trifasciata* (Forskål), banded garden spider

Description. This species has a whitish to pale yellow series of lateral stripes along the dorsum of the abdomen. The ground color is pale yellow, with silvery hairs on the carapace and with numerous thin silver and yellow transverse lines alternating with black on the abdomen. The legs are spotted. The abdomen is usually more pointed behind than in *A. aurantia,* and lacks the notch and humps in front that are characteristic of that species.

Length of the female 15 to 25 mm; length of the male 4 to 6 mm.

Biology. Females may spin one or two eggsacs before they die within a few weeks of mating. The eggsac is a brown, flat-topped, cup-shaped object, sometimes described to be the shape of a kettle drum. The eggsac is about 18 mm in diameter and may contain over 100 eggs. The shape of the eggsac can be used to indicate the species.

Range. Widespread in Texas. Reported throughout the United States.

Habitat. The webs are built in tall grass in sunny areas.

Outdated and Unofficial Names. Banded garden spider.

References. Kaston (1978). Levi et al. (1990). Platnick (1993). Breene et al. (1993b).

21J. *Cyclosa turbinata* (Walckenaer), cyclosa orbweaver

Description. Females have a pair of anterior dorsal humps with white pointed tubercles at the end of their abdomens and dark markings on the abdomen. The carapace is brownish.

Length of the female 3.3 to 5.2 mm; length of the male 2.1 to 3.2 mm.

Biology. When in cotton, the web is typically constructed toward the middle of the plant. Its prey is caught in the orb web, then wrapped. The prey carcass is hung on a vertical line radiating out from the web's center vertically instead of being discarded after consumption. Eggsacs are formed along this line. The web is renewed daily, leaving the eggsacs and prey carcasses intact (Levi, 1977). The spider normally stays at or near the center of the web and appears nearly identical to its wrapped prey.

It has been reported to feed on aphids and cotton fleahoppers (Nyffeler et al. 1986a, Breene et al. 1989a).

Range. Widespread in Texas and found from June through September.

Habitat. This small orbweaver can be quite common on cotton.

References. Breene et al. (1993b). Levi (1977a).

21K. *Eriophora ravilla* (C. L. Koch)

Description. This species is distinctive in having a long, band-like scape on the female epigynum and a more or less distinct hump on the "shoulder" of either side of the top of the anterior abdomen. Coloration is highly variable. The carapace is typically red-brown with white hairs, and the dorsum of the abdomen is white to dark gray, or brown to occasionally black.

Length of the female 12 to 24 mm; length of the male 9 to 13 mm.

Biology. The spider remains suspended head down on the web at night and spends the day hidden in partly rolled leaves.

Range. South and east Texas. Southeastern states.

Habitat. The species prefers an open woodland habitat, where it produces a large web after dark and removes it before dawn.

Taxonomic Status. Members of this genus have a hard, glossy integument with a single median tubercle behind and two "shoulder" humps on the abdomen which may not be very evident. The long scape on the epigynum is one of the key characters for the genus. The abdomen is generally highest above the spiracles. The venter has a median black patch.

E. edax (Blackwall) is an exceedingly variable species that also is recorded from south Texas. The abdomen may be dark with a mottling of light, light with a small diamond-shaped black spot up front, orange-tan with a white stripe along the mid-dorsum. or evenly pink all over. The venter shows a large black triangle with the apex pointing to the rear. Length of the female 12 to 16 mm; length of the male 11 to 13 mm.

Outdated and Unofficial Names. Southern orbweaver.

References. Breene et al. (1993b). Levi (1970).

21L. *Eustala anastera* (Walckenaer), humpbacked orbweaver

Description. The carapace is gray and darker at the sides. The abdomen shows a considerable variety of patterns. Most commonly, it is gray with a central triangle having scalloped edges. It has a hump above the spinnerets.

Length of the female 5.4 to 10 mm; length of the male 4 to 9.5 mm.

Biology. This species builds its web in the evening in the upper portion of a plant and removes the web by morning.

Range. Seen throughout Texas from May through September. Reported from New England to Florida and west to the Rockies.

Habitat. In low trees and among shrubs and bushes.

References. Kaston (1978). Levi et al. (1990). Platnick (1993). Breene et al. (1993b).

21M. *Gasteracantha cancriformis* (Linnaeus) spinybacked orbweaver

Description. The carapace is dark brown. The abdomen is white, orange, red, or yellow, with red pointed spurs, and dark brown oval spots above. The venter is black with small yellow spots. Females are most often seen. Males are smaller and shaped differently.

Length of the female 8 to 10 mm; length of the male 2 to 3 mm.

Biology. This species can be very common in wooded areas, especially in the fall. It is easily recognized and one of our most distinct spiders.

Range. Found in the eastern two-thirds of Texas. Reported from North Carolina south to Florida and west to California.

Habitat. In wooded areas.

Outdated and Unofficial Names. *Gasteracantha elipsoides* (Walckenaer).

References. Kaston (1978).

21N. *Gea heptagon* (HENTZ)

Description. The carapace is yellowish, with darker brown markings between the radial furrow. The legs are yellowish with brown annuli. The abdominal dorsum is pale yellow, but with darker spots and a posterior dark triangle.

Length of the female 4.5 to 6 mm; length of the male 2.5 to 4.5 mm.

Biology. The snare is built close to the ground in low vegetation with no stabilimentum. The spider sits in the center of its web. Occasionally, a sector is missing from the *lower* half of the web, which is distinct. (Another genus, *Zygiella,* makes webs with a sector missing from the *upper* half.) In one study, aphids made up about half of all insects intercepted in the webs of this species (Nyffeler et al. 1989). The spider drops from its web when frightened and changes color by darkening the light areas of the body. This color change makes it more difficult to see against the ground.

Eggsacs are flattened, ivory-colored, and typically contain from 30 to 45 eggs (Sabath 1969).

Range. Found in the eastern third of Texas. Reported from New Jersey south to Florida, west to Michigan, Kansas, and Texas, and then to California.

Habitat. The species has been collected on cotton from June to August but is relatively uncommon.

References. Kaston (1978). Breene et al. (1993b).

21O. *Larinia directa* (HENTZ)

Description. The carapace is yellow with a faint brown line extending back from the posterior median eyes. The abdomen is more than twice as long as wide. The abdomen is orange-yellow, quite variable in markings, but usually with indications of four pale red longitudinal lines. Some individuals show two rows of six or more black spots and a single black spot on the blunt cone of the abdomen overhanging the carapace in front. The specimens with spots on the abdomen also have spotted legs.

Length of the female 4.8 to 11.7 mm; length of the male 4.5 to 6.5 mm.

Range. Southern half of Texas. Reported from New Jersey south to Florida and west through the southern states to California

Habitat. In grass in sunny areas.

Outdated and Unofficial Names. *Drexelia.*

References. Kaston (1978). Platnick (1993).

21P. *Larinioides cornutus* (Clerck), furrow orbweaver

Description. The carapace is brown to gray and heavily setose (bristly). The abdomen is dorsoventrally flattened and has a folium which is grayish brown, somewhat darker than the sides, and with lighter areas within. The edges of the folium are usually entire, but sometimes broken. The venter of the abdomen is black between the genital furrow and spinnerets enclosing a white comma-shaped mark on each side.

Length of the female 6.5 to 14 mm; length of the male 4.7 to 9 mm.

Biology. In this genus, mature individuals of both sexes can be found at all seasons of the year, and the males are able to mate three or four times with each palp. Webs are spun in the evening and usually contain fewer than 20 radii, and have the spiral turns widely separated so that the entire web shows an "open" appearance. Silken retreats are built in crevices of walls, on railings, or among plants. Eggsacs are 7–10 mm in diameter and are covered with yellowish threads and hidden in the retreat. Eggsacs hold 50–250 eggs.

Range. Eastern two-thirds of Texas. Reported throughout the United States and Canada, northwest to Alaska. More common east of the Mississippi River and in the more northern states.

Habitat. These spiders are often found around houses and other man-made structures, and frequently in bushes near water.

Outdated and Unofficial Names. *Nuctenea, N. foliata* (Fourcroy).

Taxonomic Status. This genus may be confused with *Metazygia* which lacks setae on the carapace.

References. Kaston (1978). Levi (1974). Comstock (1940).

21Q. *Mecynogea lemniscata* (Walckenaer), basilica orbweaver

Description. The carapace is yellow to greenish, with a narrow black line running longitudinally through the middle and wider ones at the margins. The legs and sternum are greenish. The femora have indistinct longitudinal lines.

The median dorsal abdominal band is framed laterally by a red line enclosing orange-yellow areas. The anterior portion of the orange area is divided by a bluish black line and a transverse black mark. Posteriorly, the orange area grades into median bluish black. There is a white, wavy longitudinal line on each side of the band. Sides of the abdomen are bluish green.

Length of the female 6.3 to 9.0 mm; length of the male 4 to 6.6 mm.

Biology. The basilica orbweaver spins a dome-shaped web (without sticky silk), typically near the top of plants. It builds an irregular-shaped web or labyrinth near the domed orb, where it resides. The eggsacs are attached to each other in strings suspended in the web. The shape and placement of eggsacs are often an indicator of the species. Specimens of this species have been found paralyzed in the cells of mud dauber wasps.

Range. Eastern two-thirds of Texas. Reported from the District of Columbia south to Florida and west to Colorado.

Habitat. The species is found in woods and occasionally observed on cotton in May and June.

Outdated and Unofficial Names. *Allepeira conferta.*

References. Breene et al. (1993b). Kaston (1978). Levi (1980).

21R. *Metepeira labyrinthea* (HENTZ), LABYRINTH ORBWEAVER

Description. The carapace is brown and much lighter in the eye region. The abdomen is brown with a distinct folium of white and black areas.

Length of the female 5.5 to 6.2 mm; length of the male 4 to 4.5 mm.

Biology. The snare is a composite, consisting of a more or less vertical orb behind which is placed an irregular labyrinth resembling the snare of a theridiid. Often, a small silken tent is constructed in the mesh, and this serves as a retreat.

Range. Throughout Texas and widespread throughout the United States.

Habitat. In trees and shrubs.

References. Kaston (1978). Platnick (1993). Levi (1977b).

21S. *Micrathena sagittata* (WALCKENAER) ARROWSHAPED MICRATHENA

Description. Females have three pairs of pointed tubercles on the abdomen, with the posterior pair widely spreading to make the abdomen triangular or arrowhead-shaped in appearance when viewed from above. These tubercles are black at the points and red at the base. The middle of the abdomen is bright yellow to orange on the dorsum. In the male, the abdomen is widest at the rear, without spines, and the spinnerets are nearer to this end than to the pedicel. The abdomen of males is black, except the lateral sides which are white.

Length of the female 8 to 9 mm; length of the male 4 to 5 mm.

Biology. Eggsacs are fluffy white spheres 12 mm in diameter, usually containing about 90 eggs. Their webs are rarely more than 60 cm (2 ft) above the ground (Fitch 1963). The web has a short stabilimentum above the hub.

The species has been noted preying largely upon leafhoppers. Uetz and Biere (1980) reported a prey spectrum composed primarily of Diptera, Hymenoptera, Coleoptera, and Homoptera in nonagricultural habitats.

Range. Present in eastern, central, and southern Texas. Reported from eastern states west to Texas and Nebraska.

Habitat. Most often observed along the edges of open woods and in forests and brushy areas.

Taxonomic Status. At least one other species in this genus, *Micrathena gracilis* (Walckenaer), often occurs in Texas. However, the spines on the abdomen are nearly all the same size and thus the abdomen is not characteristically arrowhead-shaped.

References. Breene et al. (1993b). Kaston (1978).

21T. *Neoscona arabesca* (Walckenaer), Arabesque Orbweaver

Description. The colors are yellow and brown with paired black spots on the posterior half of the dorsum and lighter areas. Tibia 2 of the male is curved.

Length of the female 5 to 12.3 mm; length of the male 4.2 to 9.2 mm.

Biology. Webs are vertical. This species can be abundant in some situations. It can be seen from May through September. The eggsac is a lens-shaped case, 10 mm in diameter, and contains about 280 eggs.

Aphids and beetles are significant components of the prey in cotton fields (Nyffeler et al. 1989).

Range. Throughout Texas, with most records from the eastern half. Reported throughout the United States.

Habitat. The species is found largely in sunny, moist habitats and has been reported from cotton and soybean fields. The webs are built in tall grass and low bushes.

Taxonomic Status. *Neoscona* have heavy spines on the legs. There are several species in this genus that may be easily confused. The carapace has a longitudinal groove, which separates them from *Araneus*. The abdomen varies in shape from oval, elongate, triangular, or humped. All species have a black pattern on the underside of the abdomen which is bordered by white spots. Abdominal patterns can be used as a quick guide to some species. However, genitalia should be examined for positive identification. The epigynum of the female is fused, forming a simple tongue.

References. Kaston (1978). Levi et al. (1990). Platnick (1993). Breene et al. (1993b). Berman & Levi (1971). Culin and Yeargan (1982).

21U. *Neoscona oaxacensis* (Keyserling)

Description. This spider has a black and white to yellowish herringbone pattern on the dorsum of the somewhat narrow abdomen. This pattern and the narrowed abdomen separate it from most other spiders in the genus. However, the pattern on the abdomen varies across its range from rather distinct wavy lines of black and white to black with irregular white paired spots. There is also considerable variation in size. The epigynum of the female has a blunt scape with two small lobes underneath.

Length of the female 8.9 to 18.0 mm; length of the male 6.3 to 12.7 m.

Biology. *Colias* butterflies have been found in their webs. These spiders are occasionally abundant enough to be a nuisance.

Range. Western two-thirds of Texas, central Texas west to California, throughout Mexico to South America.

Habitat. Found in many habitats including citrus, carrot seed heads, alfalfa, shrubs, tall grass, cedars, and on bridges.

References. Berman and Levi (1971). Kaston (1978).

Key Family Characters

Wolf spiders are some of the most common spiders. Most are cryptically colored in brown, tan, or black. They have eight eyes in three rows with 4-2-2 arrangement (Figure 33). They have three claws and no cribellum.

The anterior eyes are the smallest and usually in a straight line of four. The posterior row is so strongly curved that it forms two rows of two each. The eyes in the second row are the largest. The third row of eyes may be nearly as large as the posterior medians. The eyes are all dark in color. The legs are usually scopulate and spiny.

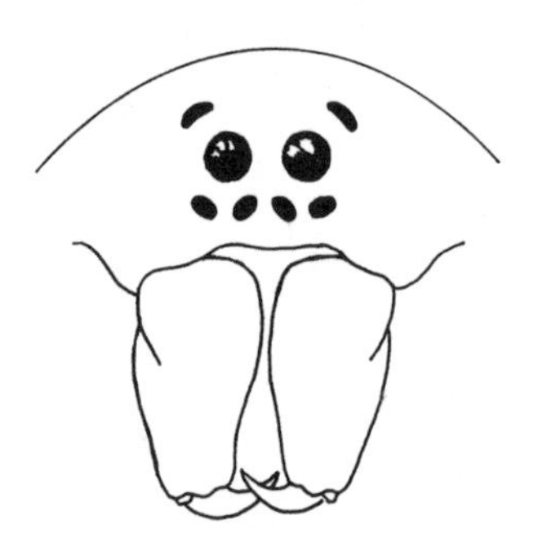

Figure 33. Front view of the carapace of a wolf spider (Lycosidae) showing eye arrangement.

Biology

Many wolf spiders are active at night but some hunt during the day. They often enter buildings. Wolf spiders usually stay on the ground but sometimes venture up on plants. They hide in sheltered areas or vegetation when not hunting, but some make burrows. They have good vision and may actively pursue prey. However, many tend to use a sit-and-wait foraging strategy more often than stalking. Some species are widespread while others are localized.

Males wave their large pedipalps rhythmically while courting females. The eggsac is globular, and may have two halves with a seam around its "equator." It is carried by the female attached to her spinnerets, except in *Sosippus*. Females attach their large eggsacs to their spinnerets, carrying them until after hatching; they will search for a lost eggsac and reattach it if found. They carry their young on their abdomens for some time after hatching.

The lycosids, pisaurids, ctenids, and certain philodromids have a tapetum in their eyes that reflects light at night. A good way to find these spiders is by holding a flashlight out from your forehead close to the eyes or by wearing a headlamp on your forehead (Whitcomb et al. 1963). A sharp pinpoint of greenish light may then be seen emanating from the eyes of the spiders and roaming over the ground or vegetation, often over great distances (40 m or more).

The large wolf spiders prey upon a wide variety of arthropod species, including some hard-bodied insects and other spiders (Kuenzler 1958, Whitcomb et al. 1963, Nyffeler et al. 1986a, Hayes and Lockley 1990). Predation on noctuid moths has been observed by Whitcomb et al. (1963).

Snares are not built, except in *Sosippus* and the immatures of some other genera. Some wolf spiders make tubular tunnels in the ground, others make use of natural depressions under rocks or debris, and still others never use a retreat, but simply run through grass, among dead leaves on the forest floor, or over sandy and stony areas.

Taxonomic Status

Perhaps more than one hundred species occur north of Mexico. This family has had significant taxonomic revisions recently. The wolf spiders in the genera *Hogna, Rabidosa,* and *Varacosa* were previously included in the genus *Lycosa*. However, some authorities now consider that the genus *Lycosa* is restricted to the old world. While the family can be easily recognized by the eye pattern, recognizing genera is quite challenging. Recent identification keys rely heavily on the genitalia of males and females.

Descriptions are in Gertsch (1934), Gertsch and Wallace (1935), Kaston (1948, 1978), Vogel (1970b), Wallace and Exline (1978), and Dondale and Redner (1978a, 1983, 1984). Further information on predation is contained in Yeargan (1975), Nyffeler and Benz (1988), and Hayes and Lockley (1990).

References

Kaston (1978). Levi et al. (1990). Breene et al. (1993b). Roth (1993).

Generic Summary for Lycosidae in Texas

Quick recognition characters	Genus	Number of species
Often pale-colored forms in sandy locations	*Arctosa*	2
Thin-legged common species	*Pardosa*	11
Carapace nearly flat in front and angling sharply at the rear. The upper surface has a dark Y shape or tuning fork shape on a light background. Small species often found in damp areas.	*Pirata*	6
Large species	*Hogna*	9
	Rabidosa	2
Burrowing species seldom seen	*Geolycosa*	5
Other genera not easily recognized		
	Allocosa	7
	Alopecosa	1
	Gladicosa	3
	Hesperocosa	1
	Isohogna	2
	Schizocosa	8
	Sosippus	1
	Trochosa	3
	Varacosa	3

22A. *Arctosa littoralis* (Hentz)

Description. This species is spotted gray or dirty white, like the color of the sand over which it runs. It blends so well that the spider is difficult to see unless it moves. It belongs to that section of the genus in which the carapace is not so smooth and the anterior row of eyes is shorter than the second.

Length of the female 11 to 15 mm; length of the male also about 11 to 15 mm.

Biology. *Arctosa* are not common but they are widespread. The body color may match the background soil on which they run.

Range. Widespread in Texas. Reported throughout the United States and eastern Canada.

Habitat. Sandy soils.

References. Kaston (1978). Levi et al. (1990). Platnick (1993).

22B. *Geolycosa,* BURROWING WOLF SPIDERS

Description. These spiders are sandy gray to brown with black markings.

Length of adults 14 to 22 mm.

Biology. Almost their entire existence is spent in their burrows, where they wait at the mouth for prey. They are sensitive to the slightest vibrations of the ground and scurry beneath at the first sign of danger. Once you have disturbed one of these spiders, it is necessary to remain motionless near the burrow for several minutes to see the spider when it returns to the surface.

These spiders dig in sand almost straight down up to one meter. They dig with their chelicerae, and the grains of sand are stuck together with silk and thrown out of the burrow. The front legs are heavy and strong, fitted for digging. The colored sand from different depths may form concentric rings around the hole as the spider removes sand from the burrow. In fair weather, the eggsac is brought to the surface and sunned. At night, the spiders stay close to the surface.

Range. There are scattered records for this genus from central and south Texas.

Habitat. Found in soils that are soft enough to dig.

References. Kaston (1978). Levi et al. (1990).

22C. *Hogna carolinensis* (Walckenaer)

Description. The carapace is dark brown with gray hairs (lighter in the males) and usually without distinct markings. The abdomen is likewise brown with a somewhat darker median longitudinal stripe. The sternum, coxae, and venter are all black.

Length of the female 22 to 35 mm; length of the male 18 to 20 mm.

Biology. This is our largest wolf spider. This species usually builds a burrow in the ground.

Range. Throughout the United States.

Habitat. Found in open country on dry hillsides and prairies.

Outdated and Unofficial Names. *Lycosa.*

References. Kaston (1978). Levi et al. (1990). Gertsch (1979).

22D. *Pardosa,* THINLEGGED WOLF SPIDERS

Description. The legs are relatively long, with the metatarsi and tarsi quite thin, and with very long spines. The tibia plus the fourth patella is usually longer than the carapace, and the first tibia has three pairs of spines of which the distal pair is apical and shorter than the others, which are very much longer than the thickness of the segment. The cephalothorax is highest in the head region, and the chelicerae are much smaller than in most other lycosids so that their length is less than the height of the head. Length is 4 to 9.5 mm.

Pardosa is a large group of spiders that are difficult to distinguish from one another.

Biology. The eggsac is lenticular, usually greenish when fresh and changing to dirty gray when older.

These spiders are commonly found in pitfall traps but venture up on plants to hunt. *Pardosa* species are usually considered to be diurnal but some species have been observed to be active most of the night (Breene et al. 1989a). *Pardosa* feed on small prey from various insect orders, including aphids (Nyffeler and Benz 1981 and 1988, Dean et al. 1987, Nyffeler and Breene 1990).

Range. Throughout Texas.

Habitat. Some species are common in cotton fields.

References. Kaston (1978). Breene et al. (1993b).

22E. *Pardosa delicatula* Gertsch & Wallace

Description. The abdomen is dull yellow in the middle and darker on the sides.

Length of the female 5 to 6.5 mm; length of the male 4.5 to 5.1 mm.

Biology. This spider consumes mosquito larvae in still-water conditions (Breene et al. 1988a), as do other *Pardosa* (Greenstone 1978, 1979a,b, 1980).

Pardosa pauxilla Montgomery is very similar in appearance and habits.

Range. Widespread in Texas from May through September.

Habitat. Common in pastures and grasslands. *P. delicatula* is at least partly aquatic. It is sometimes seen in cotton and sweetpotato fields.

References. Breene et al. (1993b). Dondale and Redner (1984).

22F. *Pardosa milvina* (Hentz)

Description. The dorsal stripes on the carapace undulate more than in the other species. This species has yellow spots on the abdomen. The males have white hairs on the palpal patella as do males of *P. atlantica.*

Length of the female 5.1 to 6.4 mm; length of the male 4.3 to 5.0 mm.

Biology. The species normally stays near the ground during the day, but relatively large numbers have been observed foraging on cotton plants at night. It has been shown to feed on cotton fleahopper and insect eggs (Breene et al. 1989a and 1989b, Hayes and Lockley 1990, Nyffeler et al. 1990).

At least two eggsacs are produced per season. The eggsacs are about 3.5 to 4.7 mm in diameter and contain about 32 to 93 eggs.

Range. Found in the eastern third of Texas from May through September. Reported from New England and adjacent Canada south to Florida and west to the Rockies.

Habitat. Found in dry open woods, as well as on wet ground and along the edges of ponds and streams. It is also reported from various agricultural crops.

References. Kaston (1978). Platnick (1993). Breene et al. (1993b). Breene et al. (1989a, 1989b) Nyffeler et al. (1990). Dondale and Redner (1984).

22G. *Pirata*, PIRATE WOLF SPIDERS

Description. The wolf spider genus *Pirata* can be distinguished by the darkened Y-shaped pattern (like a tuning fork) on a yellow band that runs dorsally on the cephalothorax from the eye region to the posterior. On either side of the median light area, the carapace is dark brown, gray, or black to a marginal or submarginal light band. The abdomen usually has a light longitudinal stripe on the anterior half and often indistinct chevrons behind. Generally, there are paired yellow spots or spots of white scales on the posterior half of the abdomen. The thoracic part is as high as the cephalic or even slightly higher. *Pirata* are our smallest species of lycosids.

Biology. Many if not all members of the genus can run across the water's surface and temporarily duck underwater to capture prey or to hide when startled. At least some of the species can prey upon mosquito larvae beneath the surface of still water (Breene et al. 1988b).

Range. Throughout Texas.

Habitat. This genus is normally associated with aquatic or semi-aquatic freshwater ecosystems. Those captured in cotton fields are probably near ponds or streams or perhaps migrating.

Taxonomic Status. *Pirata davisi* Wallace & Exline is a common species found in the eastern half of Texas.

References. Kaston (1978). Breene et al. (1993b).

22H. *Rabidosa rabida* (WALCKENAER)

Description. *R. rabida* is perhaps the most common and best known of the wolf spiders in the United States. The dorsal abdomen has a fairly distinct pattern in the form of lighter longitudinal stripes with a series of light chevrons on a darker background. The male has the first leg dark brown or black, and the venter is not spotted.

Length of the female 16 to 21 mm; length of the male about 12 mm.

Biology. Eggsacs are from 7 to 10 mm in diameter and contain from 168 to 365 eggs.

Range. Typically found in east Texas from May through September, they have been recorded in many areas of Texas. Reported from New England to Florida and west to Oklahoma and Nebraska.

Habitat. Found in wooded areas, cotton fields, buildings, and homes.

Reference. Kaston (1978). Breene et al. (1993b).

23 PISAURIDAE—NURSERY WEB SPIDERS

Key Family Characters

The nursery web spiders are large, conspicuous and often boldly marked. They resemble large brown wolf spiders, but have eight eyes of nearly equal size. They also have three tarsal claws and lack a cribellum.

The anterior eye row is straight, and the posterior eye row is so strongly recurved that it appears to form two rows. The posterior median eyes of pisaurids are about the same size as the posterior lateral eyes, but they are much more widely separated than eyes of lycosids.

Biology

These spiders are usually associated with water but may be found on vegetation or around boat docks. Pisaurids may hunt actively during the day, apparently with good vision, or they may sit quietly for hours.

They primarily prefer aquatic habitats. Many have adaptations that allow them to skate on the surface of the water and dive beneath it to search for prey or escape from enemies.

Females carry huge eggsacs in their jaws. Before eggs hatch, the female may suspend the eggsac between leaves in a shelter that she erects. She guards the eggs and young. Young spiders leave the nest in about a week.

A snare is built only by some immatures. As the common name implies, the female makes a loose shelter for the eggsac and immatures.

Taxonomic Status

Only about five genera and 15 species are known from North America north of Mexico. Carico (1973) completed a revision of the genus *Dolomedes*.

References

Levi et al. (1990). Platnick (1993). Breene et al. (1993). (Roth 1993).

23A. *Dolomedes tenebrosus* Hentz

Description. The colors are irregular brownish gray with some white mixed in.

Length of the female 15 to 26 mm; length of the male 7 to 13 mm.

Biology. Like other *Dolomedes,* this spider is usually found in aquatic habitats. However, it wanders during the summer in search of water and may be found in wooded areas, cotton fields, or other locations. Some of the larger species of *Dolomedes* are considered minor nuisances at fisheries because they capture small fish.

Range. East Texas with records from Erath and Wichita counties. Reported from New England and adjacent Canada south to Florida and west to North Dakota and Texas.

Habitat. Often found some distance from water, in wooded areas.

Taxonomic Status. *Dolomedes* are our largest species of pisaurids. There are several species, some of which are difficult to distinguish. *Dolomedes triton* (Walckenaer), the sixspotted fishing spider, is a large conspicuous species that is widespread. The carapace is gray to brown with light submarginal areas and light spots on a brown abdomen. Immatures consume mosquito larvae and other aquatic prey (Breene et al. 1988b). Length of the female 17 to 20 mm; length of the male 9 to 13 mm.

Dolomedes scriptus Hentz is similar to *tenebrosus* but has a more pronounced W-shaped white mark on the abdomen and is found primarily in the southern half of Texas.

References. Kaston (1978). Platnick (1993).

23B. *Pisaurina mira* (Walckenaer), nursery web spider

Description. The basic color is yellow to light brown with a darker brown broad median band which is margined with white on the abdomen. Another variety has the broad band less distinct.

Although all pisaurids are nursery web spiders, the specific common name of this particular species is "nursery web spider."

Length of the female 12.5 to 16.5 mm; length of the male 10.5 to 15 mm.

Biology. *Pisaurina* have no permanent homes but hunt in grass, meadows, and moist, open woods. The male presents the female with a fly during courtship. The female feeds on the fly during mating if she accepts the offering.

Range. Eastern half of Texas, with records from Kerr and Upshur counties. Reported from New England and adjacent Canada south to Florida, west to Texas, Nebraska, and Wisconsin.

Habitat. In tall grass and on bushes.

References. Kaston (1978). Levi et al. (1990). Platnick (1993).

AGELENIDAE—FUNNEL WEAVERS

Key Family Characters

Our species have eight eyes, usually similar in size, in two rows. The eye rows may be strongly recurved. They are generally dull brown in color but may be striped. The legs are long and spiny. The carapace is broadly rounded at the back and has a rectangular extension on the front. They have three claws and lack a cribellum. Spinnerets are rather long and widely separated. Males usually have coiled portions on the genitalia of the palp.

Biology

Funnel weavers are most easily located by their webs. Most of our species spin sheet or platform-like webs with a tube or funnel leading off from the center of one edge. Webs are obvious in grassy or rocky areas on dewy early mornings. The webs can nearly cover the entire surface in some locations. The spiders hide in the small end of the funnel and dart out to capture prey when it falls on the web. They pierce the prey and return to hiding to feed on it. Their rapid movements and skill at hiding make them difficult to collect.

Females deposit disc-shaped eggsacs in crevices in the fall and die soon after. Sometimes a female will still be attached to the eggsac.

Taxonomic Status

There are 25 genera and perhaps 300 species in North America north of Mexico.

References

Levi et al. (1990). (Roth 1993).

24A. *Agelenopsis,* GRASS SPIDERS

Description. Members of this genus are the largest in the family in our area. The genus can easily be recognized by the eye arrangement and by the long hind spinnerets with the distal segment about twice as long as the basal.

Generally, the carapace is yellowish to brown, with a pair of wide dark bands extending back from the lateral eyes and a thin dark marginal line on each side. The sternum is yellow to brown, often with a V-shaped dark mark. The abdomen is yellowish gray to reddish brown above with a lighter median band on the dorsum. On the venter is a broad gray band whose edges are quite dark so that in many cases there appears to be a pair of thin black lines with a light gray area between. The legs are marked with indistinct annuli, which are usually darkest at the distal ends of the segments.

Adults are usually 9 to 20 mm in length.

Biology. Webs may reach up to three feet in width and are usually built in grass, on bushes, on stone fences, or in corners of buildings, firmly attached to the substratum. Besides the horizontal sheet with its funnel retreat, the spider builds an irregular network or labyrinth which may extend far above the sheet proper. Neither the sheet nor the labyrinth are composed of adhesive threads, but the network serves to impede the flight of insects, causing them to fall upon the sheet. The spider depends upon its lightning-speed movements to capture prey.

Habitat. Commonly found in open fields and among stones of roadside fences.

Taxonomic Status. There are several very common species in this genus, all resembling one another closely. They differ principally in genitalia characters, and somewhat in size and depth of pattern pigmentation, though even in the same species there is much variation.

References. Kaston (1978). Levi et al. (1990).

24B. *Agelenopsis naevia* (Walckenaer)

Description. This is the largest, darkest, and probably most common species of the genus in our area. The abdomen is often dark chestnut brown with the markings obscure. The V-shaped mark on the sternum and the broad median band on the venter are usually not distinct, but the annuli on the legs are quite distinct.

Length of the female 16 to 20 mm; length of the male 13 to 18 mm.

Biology. Commonly found in open fields and among stones of roadside fences. Its webs may reach up to three feet in width.

Range. Records are from nearly all areas across the state. Reported from New England south to Florida and west to Kansas and Texas.

Habitat. Usually found in grassy areas or in low bushes.

Taxonomic Status. Sometimes placed in *Agelena.*

Outdated and Unofficial Names. *Agelena naevia, A. noevia.*

References. Kaston (1978). Levi et al. (1990).

24c. *Tegenaria domestica* (Clerck), barn funnel weaver

Description. The carapace is pale yellow with two gray stripes faintly indicated, and the abdomen has a number of irregular gray spots. The legs are long and faintly annulate.

Length of the female 7.5 to 11.5 mm; length of the male 6 to 9 mm.

Biology. Mature specimens may be found in all seasons, and individuals have been known to live several years. Moreover, this is one of those species in which males and females may live peaceably together on the same web during the two or three months of the breeding season.

Range. Records in Texas are from Dallas and Lubbock. Reported throughout the United States and southern Canada.

Habitat. This species is sometimes taken under stones and in rock crevices, but more often from barns, cellars, and dark corners of rooms. They frequently enter homes.

References. Kaston (1978). Levi et al. (1990). Platnick (1993).

25 HAHNIIDAE—HAHNIID SPIDERS

Key Family Characters

This is a small group of spiders with less than 20 species recorded north of Mexico. All are small in size, under 4 mm. The six spinnerets are arranged into a transverse row rather than a cluster like most spiders. They have eight eyes arranged in two slightly recurved rows. Hahniids have three claws and no claw tufts. A single broad spiracle is located well in advance of the spinnerets.

Biology

Although they are found in many habitats including forests, meadows, and cotton fields (Dean et al. 1982), these spiders are not commonly collected.

Hahniids build their delicate sheet webs (rarely more than 5 cm across) on the soil surface in small depressions. They make webs in moss, animal footprints, or moist soil. The webs are difficult to see unless they are covered with dew. The spiders live beneath grains of sand at the edge of their webs and take refuge in crevices of soil, moss, or debris.

Eggsacs of *Neoantistea agilis* (Keyserling) are circular mounds about 4 mm across and 2 mm high and covered by white silk. One reared brood had 17 spiderlings.

Prey of these spiders is not known.

Taxonomic Status

A family revision was published by Opell and Beatty (1976).

References

Kaston (1978). Breene et al. (1993b). (Roth 1993). Opell and Beatty (1976).

26 DICTYNIDAE—MESHWEB WEAVERS

Key Family Characters

Dictynids are generally small, nondescript spiders under 5 mm. They are usually brownish, grayish, or green in color. They have three claws and a cribellum.

Dictynids are typically smaller than amaurobiids with which they may be confused. The other structural characters that define the family are more difficult to see and involve the trochanters, calamistrum, and trichobothria (Roth 1993).

Biology

Webs are placed at the tips of plants, under leaves, or in crevices. Some dictynids make irregular mesh webs on plants. Female dictynids produce multiple snowy white lens-shaped eggsacs that are suspended in webbing, each containing just a few eggs. Dictynids, like uloborids, use the calamistrum to comb out silk from a sieve-like plate just forward of the spinnerets called the cribellum.

Flies, aphids, and small bugs are recorded as food (Nyffeler et al. 1994).

Taxonomic Status

They have a worldwide distribution and are the largest cribellate family, with about 350 species.

References

Kaston (1978). Breene et al. (1993b). Chamberlin and Gertsch (1958). Roth (1993).

Generic Summary for Dictynidae in Texas

Quick recognition characters	Genus	Number of species
Cephalothorax and legs usually yellowish-orange, abdomen white to gray with spots or chevrons. Some cave forms have 6 or 0 eyes and are pale colored.	*Cicurina*	69
Males with large chelicerae, concave in front and bowed outward	*Dictyna*	14
	Emblyna	12
Anterior median eyes small or missing	*Lathys*	2
Other genera not easily recognized		
	Argennina	1
	Brommella	1
	Mallos	1
	Phantyna	4
	Thallumetus	1
	Tivyna	1
	Tricholathys	1

26A. *Cicurina*

Description. *Cicurina* all have more or less the same general appearance, with the cephalothorax and legs yellowish-orange to brown, and the abdomen white to light gray with darker markings as spots or chevrons. They have eight eyes essentially in two rows as the standard, but there are species with six eyes and some which lack eyes completely. The cribellum has been reduced to a colulus. Cave forms, especially, are pallid or whitish and show a reduction in the eye pattern. Leg formula is 4123; the last leg is the longest. There are few coloration or somatic features that allow identification.

Body length is 1.2 to 7 mm with the average under 3 mm.

Biology. *Cicurina* are basically sedentary. They spin funnel webs with tangled lines of dry silk. They are found in and under surface debris and in ground openings and caves.

Range. Most of the species in Texas are associated with the caves in the Edwards Plateau regions. The genus is primarily found in the western states.

Habitat. They are generally found under stones and dead leaves on the ground. There are a number of species from Texas caves which have only six eyes or lack eyes completely.

Taxonomic Status. This genus was formerly placed in Agelenidae. Gertsch (1992) revised a subgenus *Cicurella* and described about 50 species.

References. Kaston (1978). Gertsch (1992).

26B. *Dictyna*

Description. Chelicerae of the males are large, concave in front, and bowed outward near the middle of their length. Chelicerae also have a more or less developed mastidion, (enlarged bulge at the front of the base). The calamistrum runs one-half to two-thirds the length of the middle portion of the fourth metatarsus. Color patterns are variable within a species, and related species may be similar in color. Males are usually darker than females. They are usually under 4 mm in length.

Biology. *Dictyna volucripes* Keyserling was reported as the primary spider predator on guar midges (Rogers and Horner 1977).

Range. Cosmopolitan.

Habitat. Snares are built on the ends of grass and weeds and sometimes on walls, fences, or buildings.

Taxonomic Status. Species now placed in *Emblyna* were previously included in this genus.

References. Kaston (1978). Rogers and Horner (1977).

26C. *Emblyna sublata* (Hentz)

Description. The cephalothorax of the female is yellow in front and brown in the thoracic area. The abdomen usually has an irregular broad median yellow band bordered by brown. The legs are yellow. The male has less contrast in color between the cephalic and thoracic areas of the carapace. The legs and abdomen are darker in the male than in the female.

Length of the female 2.3 to 3.7 mm; length of the male 2 to 2.5 mm.

Biology. Webs are built on the upper surfaces of large leaves.

Range. Primarily found in east and northcentral Texas. Reported from the eastern states and adjacent Canada west to Texas and the Dakotas.

Outdated and Unofficial Names. *Dictyna.*

References. Kaston (1978). Platnick (1993).

26D. *Phantyna segregata* Gertsch & Mulaik

APEX MESH WEAVER

Description. This species has orange coloration grading to dark on the sides of the carapace, and the abdomen has whitish and gray markings.

Length of the female 2.6 mm; length of the male 2.5 mm.

Biology. This is the most common species of dictynid spiders in cotton. Its mesh web is often observed in the terminals of cotton; however, Whitcomb et al. (1963) noted that webs were also built close to the ground, and the species is commonly captured in surface pitfall traps. They also may build webs on the tops or bottoms of leaves. Nyffeler et al. (1988a) found that the prey was made up chiefly of aphids and small dipterans in an eastern Texas cotton field. In another study, cotton fleahoppers were the frequent prey in a cotton field in central Texas (Nyffeler et al. 1992b).

Range. Found in the eastern half of Texas from May to September.

Habitat. Most studies have been in cotton fields, but the species is probably common in other habitats as well.

Outdated and Unofficial Names. *Dictyna.*

Reference. Breene et al. (1993b).

Numbers refer to families. Letters refer to genus or species in the text.

1. Purseweb Spider
Sphodros sp.

2. Cyrtaucheniid Spider
Myrmekiaphila sp.

3. Trapdoor Spider
Ummidia sp.

4. Funnelweb Spider
Euagrus sp.

5. **Tarantula**
(5a) *Aphonopelma* sp.

6. **Crevice Weaver**
(6a) *Kukulcania hibernalis* (Hentz), southern house spider, male

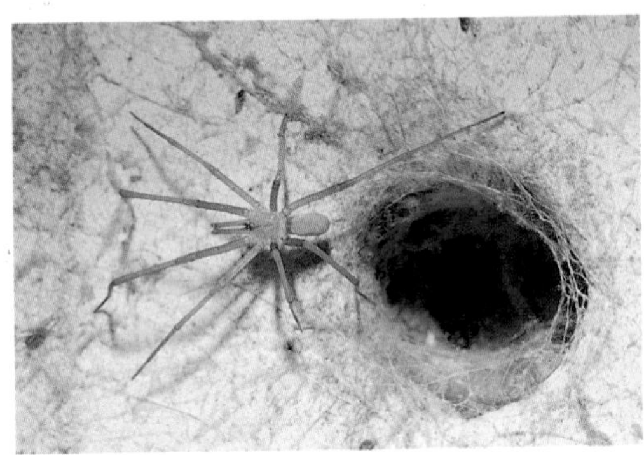

6. **Crevice Weaver**
(6a) *Kukulcania hibernalis* (Hentz), southern house spider, female

7. **Sixeyed Sicariid Spider**
(7a) *Loxosceles reclusa* Gertsch & Mulaik—brown recluse, female

8. **Spitting Spider**
Scytodes sp.

9. **Daddylongleg Spider**
(9a) *Pholcus*, possibly *P. phalangioides* (Fuesslin)—longbodied cellar spider

9. **Daddylongleg Spider**
Physocyclus sp. (with prey)

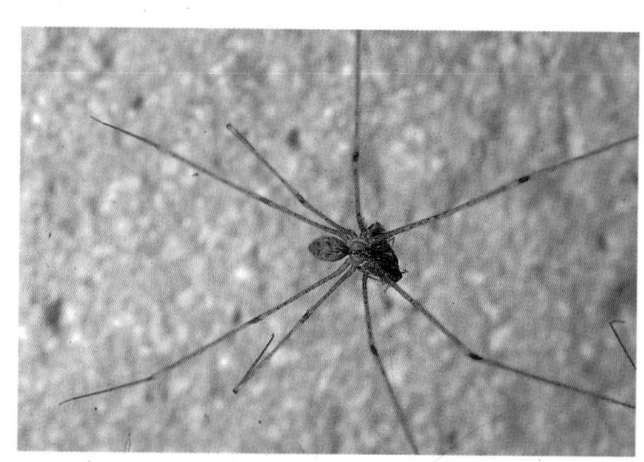

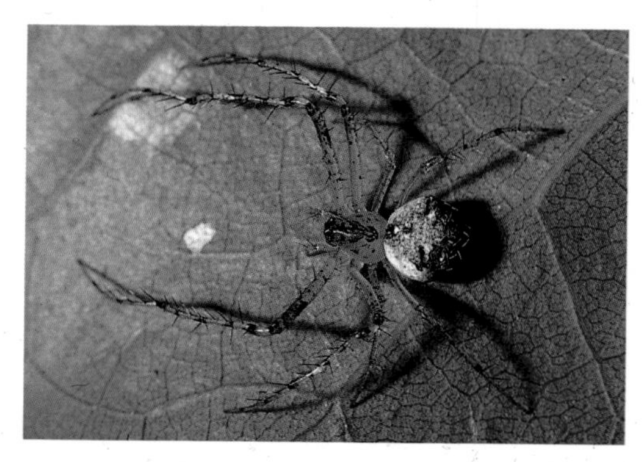

13. **Pirate Spider**
(13b) *Mimetus notius* Chamberlin

15. **Longspinneret Spider**
Tama mexicana
(O. P.-Cambridge),
male

15. **Longspinneret Spider**
Tama mexicana
(O. P.-Cambridge),
female

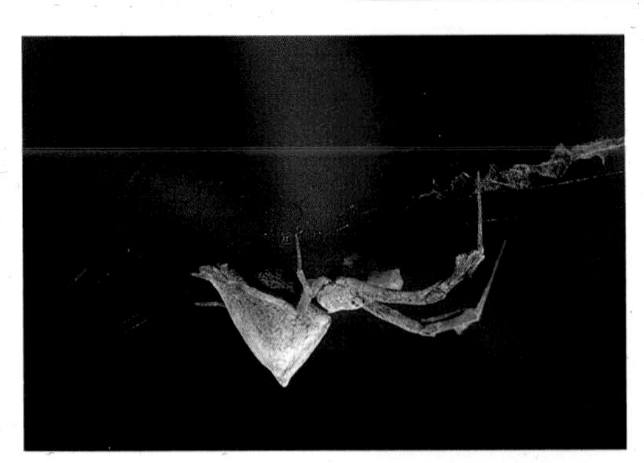

16. **Hackled Orbweaver**
(16b) *Uloborus glomosus*
(Walckenaer)—
featherlegged
orbweaver

18. **Cobweb Weaver**
(18a) *Achaearanea tepidariorum*
(C. L. Koch)—
common house
spider

18. **Cobweb Weaver**
(18c) *Argyrodes*, possibly *A. trigonum* (Hentz)

18. **Cobweb Weaver**
(18d) *Latrodectus* sp.—widow spider, male

18. **Cobweb Weaver**
(18d) *Latrodectus* sp.—widow spider, immatures

18. **Cobweb Weaver**
(18d) *Latrodectus* sp.—widow spider, eggsacs

18. **Cobweb Weaver** (18d) *Latrodectus* sp.—widow spider, female

18. **Cobweb Weaver** (18d, 18f) *Latrodectus mactans* (Fabricius)—southern black widow, immature

18. **Cobweb Weaver** (18i) *Steatoda triangulosa* (Walckenaer)

19. **Sheetweb Weaver** (19a) *Frontinella communis* Hentz—bowl and doily weaver

20a. **Longjawed Orbweaver**
(20a) *Leucauge venusta* (Walckenaer)—orchard orbweaver

P. J. Reynolds

20. **Longjawed Orbweaver**
(20b) *Nephila clavipes* (Linnaeus)—golden silk orbweaver

20. **Longjawed Orbweaver**
(20c) *Tetragnatha* sp.

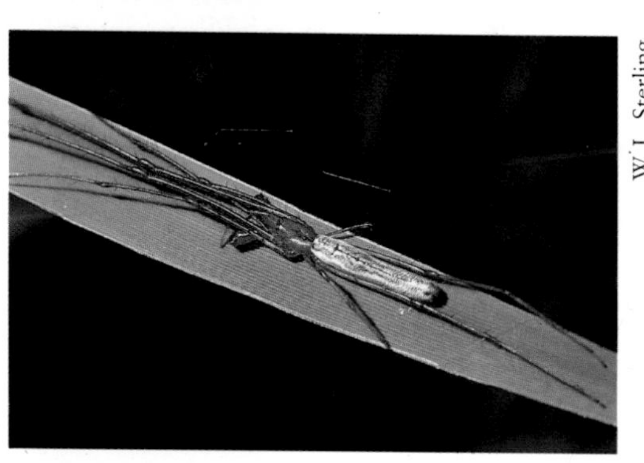
W. L. Sterling

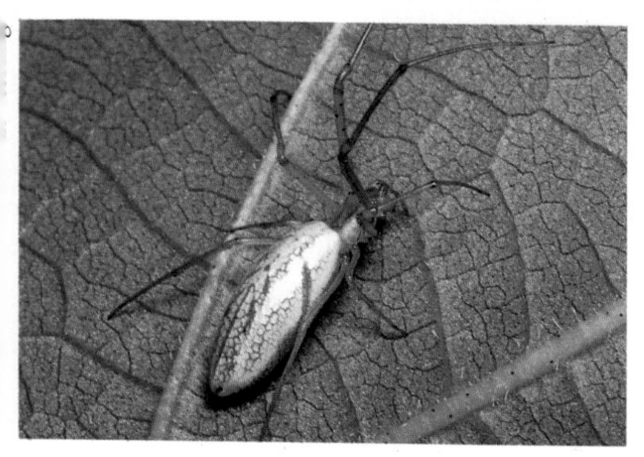

20. **Longjawed Orbweaver**
(20d) *Tetragnatha laboriosa* Hentz—silver longjawed orbweaver

21. **Orbweaver**
(21a) *Acacesia hamata* (Hentz)

21. **Orbweaver**
(21b) *Acanthepeira stellata* (Walckenaer)—starbellied orbweaver

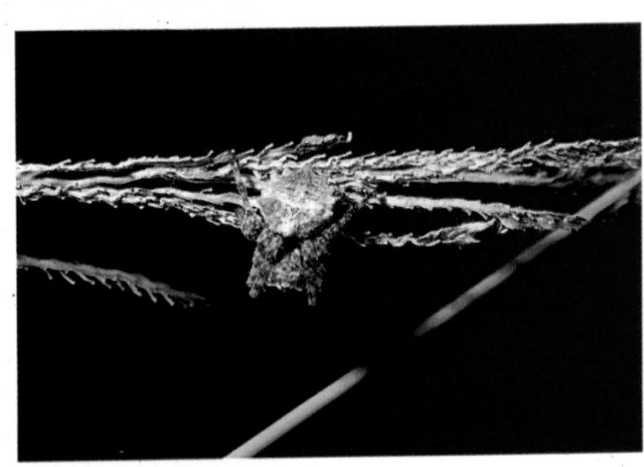

21. **Orbweaver**
(21c) *Araneus* sp.

21. **Orbweaver**
(21c) *Araneus bicentarius* (McCook) (similar sp. 21e)

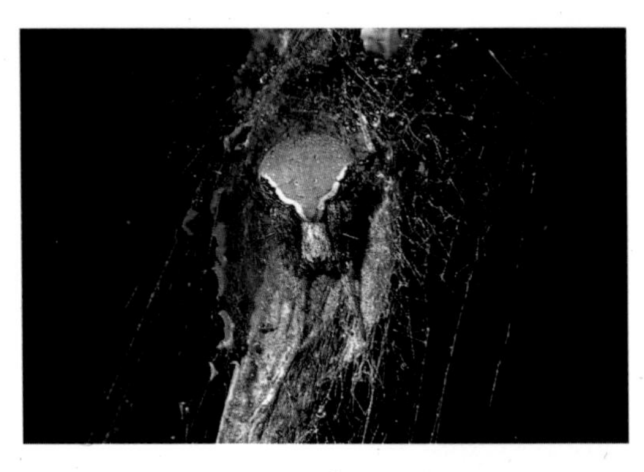

21. **Orbweaver**
(21d) *Araneus detrimentosus* (O. P.-Cambridge)

21. **Orbweaver**
(21c, 21f) *Araneus pratensis* (Emerton)

W. L. Sterling

B. M. Drees

21. **Orbweaver**
(21h) *Argiope aurantia* Lucas—yellow garden spider

21. **Orbweaver**
(21h) *Argiope aurantia* Lucas—yellow garden spider, underside

21. **Orbweaver**
(21i) *Argiope trifasciata* (Forskål)—banded garden spider

21. **Orbweaver**
(21k) *Eriophora ravilla* (C. L. Koch)

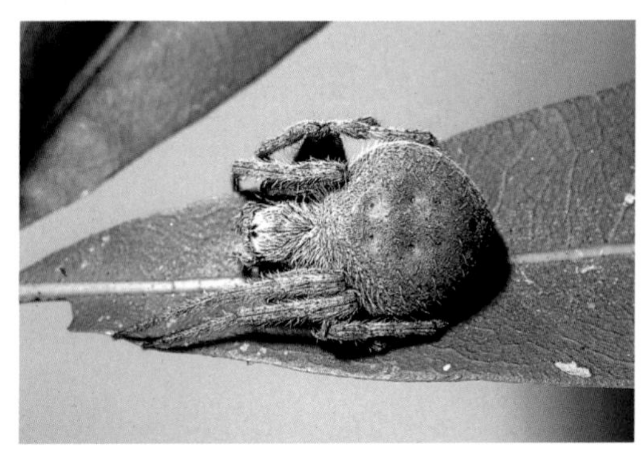

W. L. Sterling

W. L. Sterling

21. **Orbweaver**
(21l) *Eustala anastera* (Walckenaer)—humpbacked orbweaver

21. **Orbweaver**
Eustala sp. (similar sp. 21l)

21. **Orbweaver**
Eustala sp.
(similar sp. 21l)

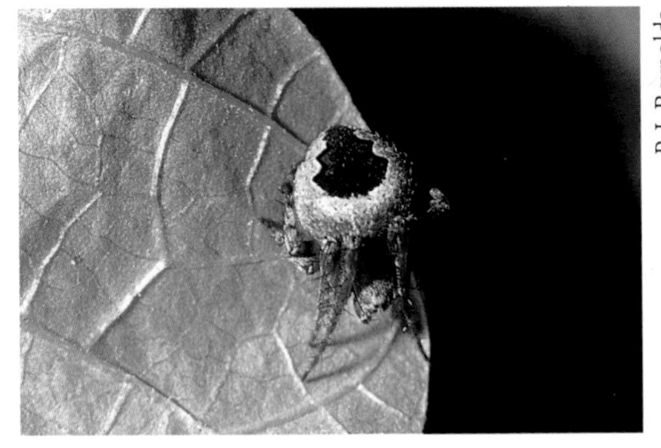

P. J. Reynolds

P. J. Reynolds

21. **Orbweaver**
(21m)
Gasteracantha cancriformis
(Linnaeus)—
spinybacked orbweaver

21. **Orbweaver**
(21m)
Gasteracantha cancriformis
(Linnaeus)—
spinybacked orbweaver

21. **Orbweaver**
(21m)
Gasteracantha cancriformis
(Linnaeus)—
spinybacked orbweaver

W. L. Sterling

21. **Orbweaver**
(21s) *Micrathena sagittata* (Walckenaer)—arrowshaped micrathena

21. **Orbweaver**
Neoscona sp. (similar spp. 21t–21u)

W. L. Sterling

21. **Orbweaver**
(21t) *Neoscona arabesca* (Walckenaer)—arabesque orbweaver

22. **Wolf Spider**
female with eggsac

22. **Wolf Spider**
(22c) *Hogna*, probably *H. carolinensis* (Walckenaer)

W. L. Sterling

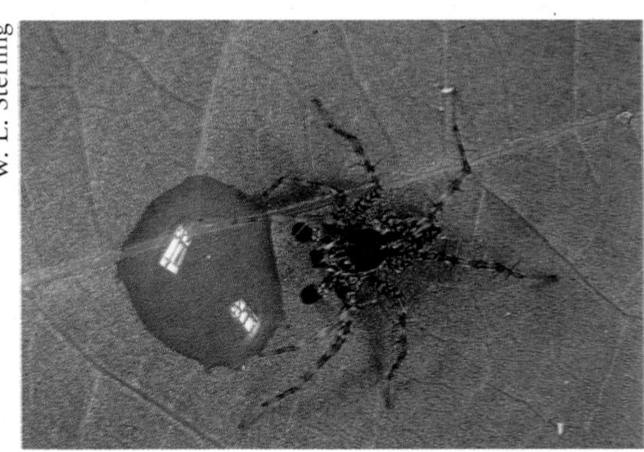

22. **Wolf Spider**
(22d) *Pardosa* sp.—thinlegged wolf spider

22. **Wolf Spider**
(22d) *Pardosa pauxilla* Montgomery (similar spp. 22e–22f)

W. L. Sterling

22. **Wolf Spider**
(22g) probably *Pirata* sp.—pirate wolf spider

22. **Wolf Spider**
Rabidosa sp.
(similar sp. 22h)

22. **Wolf Spider**
(22h) *Rabidosa rabida* (Walckenaer)

W. L. Sterling

22. **Wolf Spider**
Schizocosa avida (Walckenaer)

22. **Wolf Spider**
Varacosa acompa (Chamberlin)

W. L. Sterling

23. **Nursery Web Spider**
Dolomedes sp. or *Pisaurina* sp.

23. **Nursery Web Spider**
probably *Pisaurina dubia* (Hentz)

23. **Nursery Web Spider**
(23a) *Dolomedes tenebrosus* Hentz

P. J. Reynolds

23. **Nursery Web Spider**
(23b) *Pisaurina mira* (Walckenaer)

24. **Funnel Weaver**
(24a) *Agelenopsis* sp.—grass spider

24. **Funnel Weaver**
(24b) *Agelenopsis naevia* (Walckenaer)

W. L. Sterling

27. **Hackledmesh Weaver**
(27a) *Metaltella simoni* (Keyserling)

29. **Lynx Spider**
Hamataliwa sp.
(similar sp. 29a)

29. **Lynx Spider**
(29c) *Oxyopes salticus* Hentz—striped lynx spider, male

W. L. Sterling

W. L. Sterling

29. **Lynx Spider**
(29c) *Oxyopes salticus* Hentz—striped lynx spider, female

29. **Lynx Spider**
(29e) *Peucetia viridans* (Hentz)—green lynx spider

30. **Ghost Spider**
(30a) *Hibana gracilis* (Hentz)—garden ghost spider

33. **Sac Spider**
(33a) *Cheiracanthium inclusum* (Hentz)—agrarian sac spider

P. J. Reynolds

34. **Antmimic Spider**
(34b) *Castianeira descripta* (Hentz)—redspotted antmimic

34. **Antmimic Spider**
(34a) *Castianeira trilineata* (Hentz)

34. **Antmimic Spider**
(34c) *Trachelas* sp.

35. **Ground Spider**
(35c) *Herpyllus ecclesiasticus* Hentz—parson spider

39. **Running Crab Spider**
(39b) *Philodromus* sp.

39. **Running Crab Spider**
(39c) *Tibellus duttoni* (Hentz)

W. L. Sterling

40. **Crab Spider**
Bassaniana versicolor (Keyserling)—bark crab spider

40. **Crab Spider**
(40b) *Misumenoides formosipes* (Walckenaer)—redbanded crab spider, male

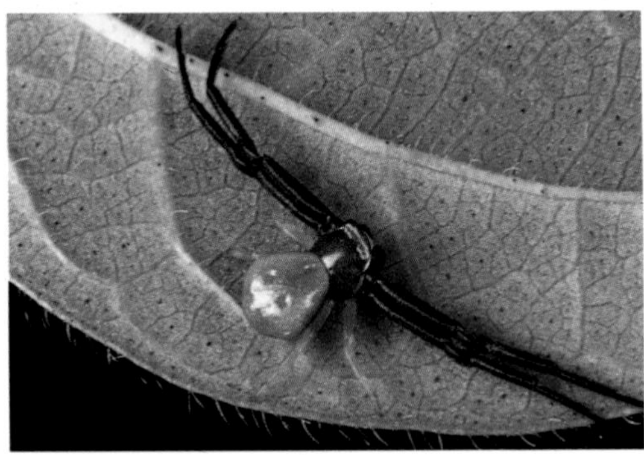

W. L. Sterling

40. **Crab Spider**
(40b) *Misumenoides formosipes* (Walckenaer)—redbanded crab spider, female

40. **Crab Spider**
Misumenops sp.
(similar spp.
40c–40e)

W. L. Sterling

W. L. Sterling

40. **Crab Spider**
Misumenops sp.
(similar spp.
40c–40e)

40. **Crab Spider**
(40c) *Misumenops asperatus*
(Hentz)—northern
crab spider

W. L. Sterling

40. **Crab Spider**
(40f) *Xysticus* sp.

W. L. Sterling

40. **Crab Spider**
(40f) *Xysticus auctificus* Keyserling

40. **Crab Spider**
(40f) *Xysticus ferox* (Hentz)

W. L. Sterling

40. **Crab Spider**
(40f) *Xysticus*, probably *X. funestus* Keyserling

41. **Jumping Spider**
(41a) *Corythalia canosa* (Walckenaer)

41. **Jumping Spider**
(41b) *Messua limbata* (Banks)

W. L. Sterling

W. L. Sterling

41. **Jumping Spider**
(41b) *Eris militaris* (Hentz)—bronze jumper, male

41. **Jumping Spider**
(41b) *Eris militaris* (Hentz)—bronze jumper, female

W. L. Sterling

41. **Jumping Spider**
(41d) *Hentzia palmarum* (Hentz) (with red mite in chelicerae)

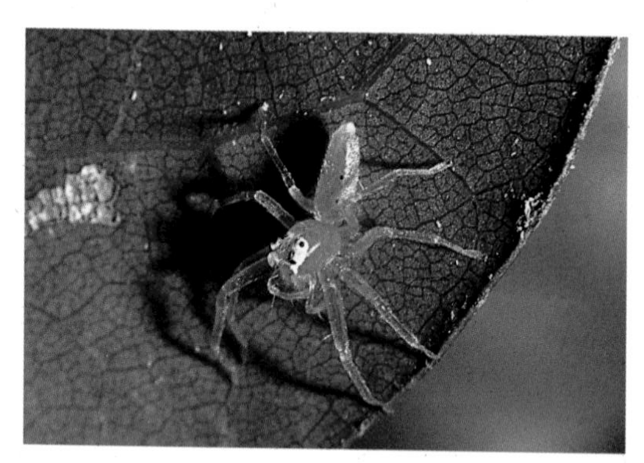

41. **Jumping Spider** (41e) *Lyssomanes viridis* (Walckenaer)—magnolia green jumper

41. **Jumping Spider** (41i) *Peckhamia picata* (Hentz)—antmimic jumper

41. **Jumping Spider** (41j) *Pelegrina*, probably *P. galathea* (Walckenaer)—peppered jumper

41. **Jumping Spider** *Pelegrina*, probably *P. exiguus* (Banks) (similar sp. 41j)

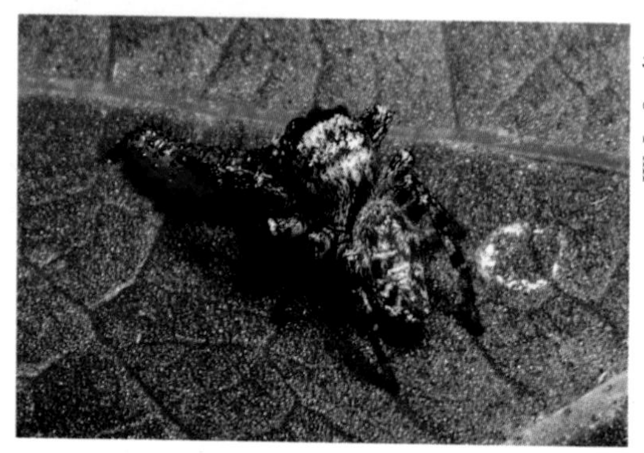

W. L. Sterling

41. **Jumping Spider**
(41k)
Phidippus sp.
(similar sp. 41m)

W. L. Sterling

W. L. Sterling

41. **Jumping Spider**
(41l) *Phidippus audax* (Hentz)—
bold jumper

41. **Jumping Spider**
Platycryptus undatus (DeGeer)
(similar sp. 41h)

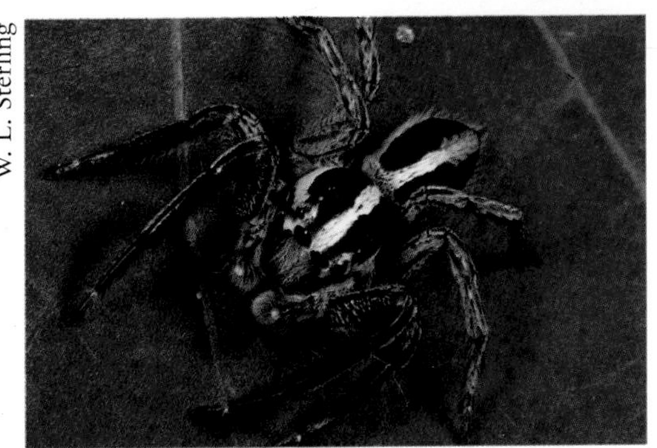
W. L. Sterling

41. **Jumping Spider**
(41o) *Plexippus paykulli* (Audouin)—
pantropical jumper, male

41. **Jumping Spider** (41o) *Plexippus paykulli* (Audouin)—pantropical jumper, female

41. **Jumping Spider** (41q) *Thiodina sylvana* (Hentz), male

41. **Jumping Spider** (41q) *Thiodina sylvana* (Hentz), female

41. **Jumping Spider** *Zygoballus* sp. (similar sp. 41r)

42. **Harvestman**
Order: Opiliones

42. **Harvestman**
Order:
Opiliones

42. **Harvestman**
Order: Opiliones

43. **Hardbacked Tick**
(43a) *Amblyomma americanum* Linn.—lone star tick, male

43. **Hardbacked Tick**
(43a) *Amblyomma americanum* Linn.—lone star tick, female

43. **Hardbacked Tick**
(43a) *Amblyomma americanum* Linn.—lone star tick, engorged female and larvae

43. **Hardbacked Tick**
(43a) *Amblyomma americanum* Linn.—lone star tick, nymphs

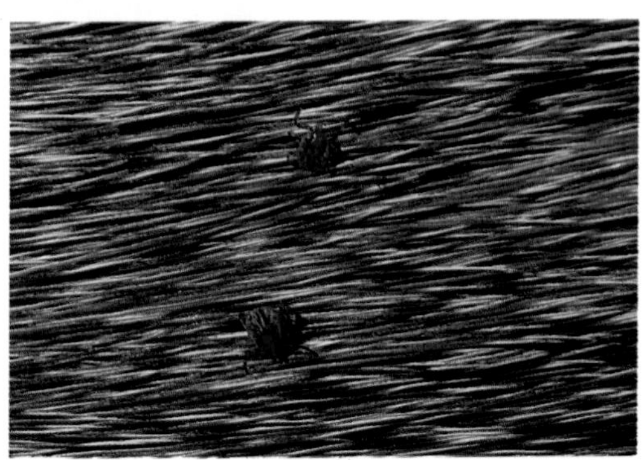

43. **Hardbacked Tick**
(43b) *Amblyomma cajennense* (Fab.)—Cayenne tick, male

43. **Hardbacked Tick**
(43b) *Amblyomma cajennense* (Fab.)—Cayenne tick, female

43. **Hardbacked Tick**
(43c) *Amblyomma maculatum* Koch—Gulf Coast tick, male

43. **Hardbacked Tick**
(43c) *Amblyomma maculatum* Koch—Gulf Coast tick, female

43. **Hardbacked Tick**
(43f) *Rhipicephalus sanguineus* (Latreille)—brown dog tick, male

43. **Hardbacked Tick**
(43f) *Rhipicephalus sanguineus* (Latreille)—brown dog tick, female

Garland McIlveen

45. **Scorpion**
Centruroides vittatus (Say)—striped bark scorpion

46. **Scorpion**
(46a) *Paruroctonus gracilior* (Hoffman)

46. **Scorpion**
(46b)
Paruroctonus utahensis
(Williams)

W. D. Sissom

46. **Scorpion**
(46c) *Vaejovis coahuilae* Williams

46. **Scorpion**
(46d) *Vaejovis reddelli*
Gertsch & Soleglad

W. D. Sissom

46. **Scorpion**
(46e) *Vaejovis waueri*
Gertsch & Soleglad

47. **Whipscorpion**
Mastigoproctus giganteus
(Lucas)—giant vinegaroon

48. **Windscorpion**
Order: Solifugae

49. **Pseudoscorpion**
Order:
Pseudoscorpiones

27 AMAUROBIIDAE—HACKLEDMESH WEAVERS

Key Family Characters

Amaurobiids are stout-legged cribellate spiders with three claws. The eyes are in two wide, nearly straight rows. The cribellum is in two parts. The calamistrum of the female is about half as long as the fourth metatarsus and is delimited by a spine on each end. The calamistrum of the male is vestigial.

Amaurobiids are usually larger than dictynids, typically 4 mm or larger. The legs usually have macrosetae and numerous tarsal trichobothria which increase in length distally. See Roth (1993) for more specific characters.

Biology

These spiders may be found in logs, under loose bark of trees, or in rock piles. They produce loose webs with coarse hackling.

Taxonomic Status

The family Amaurobiidae has traditionally been an assemblage of genera that do not fit well into other families. Recent studies have moved many genera from Amaurobiidae to other families. They are included in Dictynidae by some authors and even in the family Clubionidae by others.

References

Kaston (1978). Platnick (1993). Leech (1972). Roth (1993).

27A. *Metaltella simoni* (KEYSERLING)

Description. The carapace is orange-yellow to chestnut brown and nearly bare. It is darkened in the cephalic regions. Legs are about the same color as the carapace. Chelicerae are brown. The abdomen is covered with short setae and may be mottled gray or gray-black and somewhat darker than the carapace. In this genus, the trichobothria are long and thin. They increase in length distally on the tarsi and metatarsi.

Length of the female 8 to 9 mm; length of the male about 7 to 8.5 mm.

Biology. *M. simoni* is a nocturnal hunter. It builds small webs attached to logs and the ground. The male and female may be found together in the web.

Range. Primarily recorded from coastal and central Texas. Reported from coastal states from North Carolina to Alabama, Texas, and California.

Habitat. Often found in leaf litter, under logs, and in homes.

References. Roth (1993). Leech (1972).

28 TITANOECIDAE

Key Family Characters

Titanoecids are typically larger than dictynids and most amaurobiids. They have a cribellum and three claws. The eight eyes are homogeneous and in rows. The calamistrum usually appears biseriate (double-rowed), but may be reduced or absent in males.

Titanoeca females have the calamistrum as long as the metatarsus. *Titanoeca* have short trichobothria, which barely extend above the tarsal hairs and do not increase in length distally.

Biology

These spiders are found under loose stones and dead leaves. They make loose webs with coarse hackling.

Taxonomic Status

This group of spiders has been previously placed in the Amaurobiidae. Roth (1993) includes these in Amaurobiidae.

Reference

Leech (1972).

28A. *Titanoeca americana* EMERTON

Description. The male has the carapace a dull orange to reddish color, covered with long black hairs. The legs are orange, usually lighter distally, and the dorsum of the abdomen is gray to black. Chelicerae are reddish and darker than the carapace. The female is colored like the male except that the legs and underside are dark brown.

In the female, the first and second metatarsi show only two distal spines at the end when viewed from below. The calamistrum is composed of a single row of bristles occupying almost the entire length of the fourth metatarsus.

Length of the female 3.5 to 7.5 mm; length of the male 4.5 to 7 mm.

Biology. These spiders have been collected from April to November. Eggsacs are 7–8 mm in diameter and covered loosely with debris. Two eggsacs studied had 25 and 85 eggs.

Range. Widespread throughout Texas. Reported in New England and adjacent Canada south to Virginia and west to New Mexico.

Habitat. This species occurs under loose stones, logs, bark, and dead leaves in rather dry regions.

Taxonomic Status. This genus is sometimes placed in the Amaurobiidae.

References. Levi et al. (1990). Brignoli (1983). Platnick (1989). Leech (1972).

OXYOPIDAE—LYNX SPIDERS

Key Family Characters

These spiders are easily recognized by the arrangement of the eyes (Figure 34). They have six larger eyes that form a hexagon and two smaller eyes below in front. The legs are conspicuously spinose with the spines standing out at a considerable angle (Figure 35). The legs are long and have many long spines. The abdomen often tapers to a point behind. Lynx spiders have eight eyes, three claws, and a cribellum. They also have two rows of tarsal trichobothria. (Roth 1993).

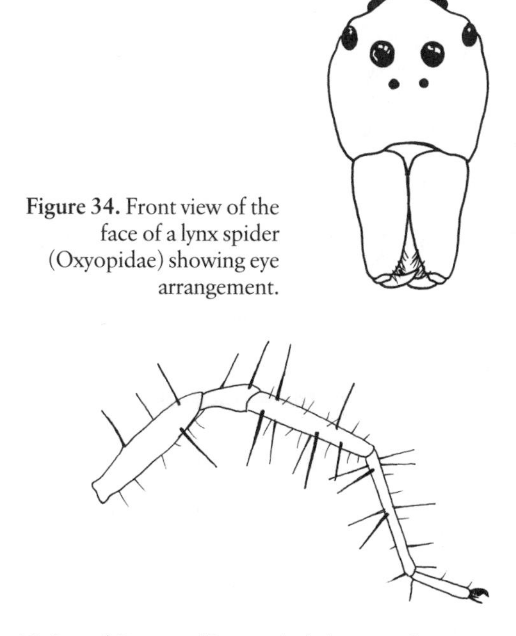

Figure 34. Front view of the face of a lynx spider (Oxyopidae) showing eye arrangement.

Figure 35. Leg of *Oxyopes* (Oxyopidae) showing characteristic spines.

Biology

Oxyopidae hunt by waiting or running rapidly and can also jump. They live among low bushes and herbaceous vegetation to which they fasten their eggsacs and over which they hunt their prey. They are day-active, and their quick, darting movements and sudden leaps are characteristic.

These spiders build no snares, retreats, or molting nests. They use silk for a drag line while hunting. They build nests for eggs, and females guard the eggs until they hatch.

The lynx spiders are probably the most economically important family of spiders in cotton ecosystems from central Texas eastward. Most live on tall grass and native vegetation that may act as a predator reservoir for continuous recolonization of cotton fields each spring (Nyffeler et al. 1992a). There is apparently one generation per year.

Taxonomic Status

Less than 20 species occur in North America north of Mexico. Brady (1964) published a revision of this family.

References

Kaston (1978). Levi et al. (1990). Breene et al. (1993b).

29A. *Hamataliwa grisea* Keyserling

Description. The carapace is orange to reddish brown with a whitish band extending back along each side in the female. These bands are lacking in the male. The dorsum of the abdomen is brown, gray, or black. The whole body is covered with fine black and white hairs which makes the spiders well camouflaged on tree bark.

Length of the female 8.7 to 10.9 mm; length of the male 8.4 to 11.1 mm.

Biology. Little is known about the biology of this species.

Range. Southern half of Texas, Brazos, and Travis counties. Reported from Florida west across the southern tier of states to California.

Habitat. On the bark of woody shrubs and trees, where the spiders blend with the background. Sometimes found on herbaceous vegetation.

References. Kaston (1978). Brady (1964).

29B. *Oxyopes apollo* Brady

Description. The face has a brown band down the front of the clypeus and chelicerae. The legs are without the ventral bands on the femora. The carapace is yellowish orange, with a brownish band on each side. The abdomen has a central light band, but is brown on the sides.

Length of the female 4.2 to 6.7 mm; length of the male 3.4 to 4.4 mm.

Biology. Not much is known about the biology of these spiders. They are sometimes captured in pitfall traps.

Range. Widespread in Texas from May through September. Reported from Tennessee, Missouri, Arkansas, Texas, and west to Arizona.

Habitat. Sometimes seen in cotton fields.

References. Kaston (1978). Breene et al. (1993b). Brady (1964).

29C. *Oxyopes salticus* HENTZ, STRIPED LYNX SPIDER

Description. The carapace is mostly yellow, with four indistinct longitudinal gray bands running lengthwise from the eyes to the pedicel. The abdomen of the female is lighter than the carapace, darker along the sides, and with a black band on the venter. The abdomen of the male is entirely gray or black above and below, with scales that make it appear iridescent. Black triangular pedipalps are very conspicuous in the front of the male. The femora of the first three legs are yellow, with a single narrow jet black line on the ventral side. There is a similar black line extending from each anterior median eye down the clypeus and the front of each chelicera. The thin, spindly legs are armed with many long spines.

Length of the female 5.7 to 6.7 mm; length of the male 4 to 4.5 mm.

Biology. This species is a hunting spider and builds no web to capture prey. It has been found to be the most abundant spider in cotton fields and other agricultural crops of the southern United States east of the Rocky Mountains (Dean & Sterling 1987). It has been documented as a predator of the cotton fleahopper (Dean et al. 1987, Breene 1988, Breene and Sterling 1988, Breene et al. 1988a, 1989a, 1989b, 1990). Striped lynx spiders also consume bollworm and tobacco budworm eggs and larvae and other prey (Young and Lockley 1985, 1986, Young and Edwards 1990, Nyffeler et al. 1987a, 1990, 1992b, 1992c). The biology of *O. salticus* has been described by Whitcomb and Eason (1967).

This species is readily captured with sweep nets. It easily disperses into other habitats by ballooning (Dean and Sterling 1990).

It has the habit of running rapidly and erratically and jumping, which is useful for identification in the field. The female attaches the disk-like eggsac to a substrate such as a leaf and guards it until the young emerge.

Range. This species is found throughout the state, but is most abundant in the eastern half of Texas. It is reported throughout the United States but is more common in the East and uncommon in the Rocky Mountains and Great Basin areas.

Habitat. In grass and field crops including cotton.

References. Kaston (1978). Breene et al. (1993b). Sterling (1982). Brady (1964).

29D. *Oxyopes scalaris* HENTZ

Description. This species is highly variable in color pattern and intensity. The carapace is mostly yellow with four longitudinal gray bands. A pair of black lines on the carapace extends from the anterior median eyes down onto the chelicerae.

The female abdomen is lighter in color than the carapace, darker at the sides, and with a black band on the underside. The male has the abdomen entirely gray or black, with iridescent scales. Femora of the first three legs are yellow with a narrow black line on the underside.

Length of the female 7 to 8 mm; length of the male 5 mm.

Biology. One egg case reportedly had 45 eggs.

Range. Throughout the United States, but less common in the central part of the country.

Habitat. Can be found in nearly any habitat. Found on trees, sagebrush, and similar shrubs, and on roadsides. Also common in agricultural fields.

References. Kaston (1978). Breene et al. (1993b). Sterling (1982). Brady (1964).

29E. *Peucetia viridans* (Hentz), GREEN LYNX SPIDER

Description. This common species has a predominantly bright green body with paler green to yellow legs. Legs are long, spindly, and equipped with black spines and spots, especially on the femora. The cephalothorax is highest in the eye region, where it is quite narrow but broadens out considerably behind. The body is bright green and usually has red spots in the eye region and over the abdomen. The red spots may be quite variable. Spiderlings are orange immediately after emergence but soon turn the familiar green.

Length of the female 14 to 16 mm; length of the male 12 to 13 mm.

Biology. These spiders lay in ambush on the blooms of wildflowers and overpower prey as they visit the flowers. They blend in well with the bright green foliage. Bees (including honeybees) constitute 23% of the prey of green lynx spiders in a cotton field. The spiders also feed on cotton fleahopper and boll weevil (Nyffeler et al. 1992c).

The green lynx spider can be common on cotton, where it may be a significant predator of cotton fleahoppers and Lepidoptera larvae and eggs (Breene et al. 1989a, Nyffeler et al. 1990). Adults perch near the top of the plant and feed upon a wide range of prey (Turner 1979, Randall 1982, Nyffeler et al. 1987c, 1992c). At times, the green lynx spider appears so fond of honeybees and other beneficial insects that at least one author (Randall 1982) questioned whether the species could be considered beneficial. Sphecid and vespid wasps, cotton leafworm larvae, bollworm adults, and boll weevil adults are also included on the prey list (Whitcomb et al. 1963, Nyffeler et al. 1992c).

These spiders typically mature about July. In the fall, adults mate while suspended in space on a drag line (Exline and Whitcomb 1965, Whitcomb and Eason 1965, Bruce and Carico 1988). Their straw-colored eggsacs are about 1.2 to 2.5 cm in diameter and contain from 129 to 602 eggs. Females guard the eggsacs (Whitcomb 1962, Whitcomb et al.1966) and are known to build foliage shelters for them (Willey and Adler 1989). They have been observed spitting venom from their fangs when disturbed while guarding the eggsacs (Fink 1984).

Range. Green lynx spiders are found throughout Texas, mainly from July through October. They are reported from North Carolina south to Florida and west to California.

Habitat. In grass, flowers, and various foliage, usually positioned at the top of the plant.

Taxonomic Status. *P. longipalpis* F. O. P.-Cambridge occurs rarely in the Lower Rio Grande Valley.

Outdated and Unofficial Names. *P. abboti* (Walckenaer).

References. Kaston (1978). Levi et al. (1990). Breene et al. (1993b). More information on green lynx spiders can be found in Kaston (1972), Weems and Whitcomb (1977), Randall (1977, 1978), Turner (1979), Killebrew (1982), Killebrew and Ford (1985), and Fink (1986).

Key Family Characters

Many anyphaenids have a pale gray color which gives them the common name ghost spiders. The ghost spiders can be differentiated from other spider families by the distinctive 5–10 lamelliform setae at the base of the tarsal claws and the tracheal spiracle which is at least halfway forward from the spinnerets to the epigastric furrow. They can easily be mistaken for clubionids or even agelenids.

Biology

Ghost spiders are usually found hunting prey on foliage. They produce small sac-like retreats similar to the retreats of clubionids.

Taxonomic Status

This family was revised by Platnick (1974).

References

Kaston (1978). Roth (1993).

30A. *Hibana gracilis* (HENTZ), GARDEN GHOST SPIDER

Description. The body is yellowish, with a darker anterior end and brown jaws (chelicerae). A pair of longitudinal gray bands is found on the top of the carapace between the eyes and the abdomen. Reddish-brown to black spots form two indistinct longitudinal bands on the abdomen. The legs are covered with spines. The tracheal spiracle is closer to the epigastric furrow than to the spinnerets, which may help to distinguish this genus from the clubionids if the lamelliform hairs cannot be seen.

Length of the female 6.4 to 8.4 mm; length of the male 5.7 to 6.5 mm.

Biology. *Hibana gracilis* builds a tube-web near the apex of cotton and other plants. It is as a predator on the cotton fleahopper (Breene et al. 1989a, b) on cotton and woolly croton. Nyffeler et al. (1990) also listed it as a predator of insect eggs. The eggsac ranges in size from 5 to 8 mm and is attached to a substrate after its construction. Eggsacs contain from 134 to 196 eggs.

Range. The species is found from May through September in the eastern half of Texas.

Habitat. On foliage, including cotton plants.

Outdated and Unofficial Names. *Aysha.*

References. Breene et al. (1993b). Kaston (1978). Platnick (1993).

31 MITURGIDAE—MITURGID SPIDERS

Key Family Characters

Miturgids, like clubionids, have two claws and eight eyes in two rows. The tracheal spiracles are near the spinnerets, and the legs are prograde. The contiguous and conical anterior spinnerets separate this genus from gnaphosids (Roth 1993).

Biology

Miturgids are nocturnal hunting spiders. No snare is built. They make retreats in litter, under rocks, or in rolled leaves.

Taxonomic Status

Several genera that were formerly in Clubionidae have now been transferred to other families. The three genera from Texas that are now considered in Miturgidae are *Strotarchus, Syspira,* and *Teminius*. Two species of *Strotarchus* have been reported from Texas. *Teminius affinis* Banks is widespread in Texas.

Outdated and Unofficial Names

Clubionidae (in part).

References

Roth (1993).

32 LIOCRANIDAE—LIOCRANID SPIDERS

Key Family Characters

Liocranids, like clubionids, have two claws and eight eyes in two rows. They have tracheal spiracles near the spinnerets. The contiguous and conical anterior spinnerets separate them from the gnaphosids. The legs have a prograde arrangement.

Biology

Liocranids are nocturnal hunting spiders. No snare is built. They make retreats in litter, under rocks, or in rolled leaves.

Taxonomic Status

Several genera that were formerly in the Clubionidae have now been transferred to other families. The characters that separate these families still seem to be unclear. The genera from Texas that are now considered in Liocranidae are *Phrurolithus, Phrurotimpus,* and *Scotinella.* Keys to genera are limited to males.

Outdated and Unofficial Names

Clubionidae (in part).

Reference

Roth (1993).

33 CLUBIONIDAE—SAC SPIDERS

Key Family Characters

This family was formerly quite large, but many of the genera are now in Miturgidae, Liocranidae, or Corinnidae. The sexes are similar, although males tend to be smaller, often with longer legs, and with longer and narrower chelicerae. Most of the species that remain in this family are pale in color.

These spiders resemble Gnaphosidae but can be distinguished by closely spaced, conical front spinnerets. They typically have a less flattened abdomen and longer legs than Gnaphosidae. Clubionids have two claws, eyes in two rows, and a prograde arrangement of the legs (Roth 1993).

Biology

Clubionids are two-clawed hunting spiders commonly encountered on foliage or on the ground. They are mostly nocturnal. They make tubular retreats in rolled up leaves, under stones, and in litter and debris.

Taxonomic Status

Several genera that were formerly in Clubionidae have now been transferred to other families. The characters that separate these families still seem unclear. The genera from Texas that are now considered in Clubionidae are *Cheiracanthium, Clubiona,* and *Elaver.*

References

Kaston (1978). Breene et al. (1993b). Edwards (1958).

33A. *Cheiracanthium inclusum* (HENTZ), AGRARIAN SAC SPIDER

Description. The body is pale yellow to pale green with dark brown chelicerae. The front leg is longer than the others. The species has a distinct lance-shaped mark on top of the abdomen. The color of the prey eaten determines the abdominal shade to some extent.

Length of the female 4.9 to 9.7 mm; length of the male 4.0 to 7.7 mm.

Biology. Males of *C. inclusum* complete 4 to 10 instars before molting into adults (mean 112 days), and most mature after the fifth or sixth stadia. Females take 5 to 10 instars to reach adulthood (mean 142 days), most maturing after the sixth or seventh instar. Laboratory-raised mature males live an average of 43 days, and females an average of 70 days. Females produce from one to five eggsacs over their life cycle, each with a mean of 38 eggs. The pale yellow, round eggs are visible within the thin, oblate spheroid eggsac. The female makes a more tightly woven brood cell and remains inside with the eggs. These spiders become aware of their prey by touching it with tarsi or palpi rather than by eyesight or web vibrations (Peck and Whitcomb 1970).

Various species of the genus have been implicated in human envenomation. The venom is considered poisonous to some extent, but not as severe as that of recluse and widow spiders. Kaston (1948) noted that the bite is no worse than the bite of a bee or wasp sting.

Range. Found throughout Texas from May through September. Reported throughout the United States except for the most northern tier of states.

Habitat. This species spins a silken tube among the leaves of shrubs, and is sometimes found in buildings.

Taxonomic Status. *C. mildei* L. Koch has not been reported from Texas. The spelling for the genus name has recently been corrected.

Outdated and Unofficial Names. *Chiracanthium.*

References. Kaston (1978). Levi et al. (1990). Platnick (1993). Breene et al. (1993b).

33B. *Clubiona*

Description. These spiders are generally white, cream, or tawny, with darker brown at the cephalic end and on the chelicerae. The body is covered with short hairs which give a silky reflection. Most species have few markings and hence are difficult to identify except by genitalia or secondary sexual characters. The chelicerae are stout in the females and more slender, long, and tapering in the males. Chelicerae sometimes have sharp ridges, or keels, along the anteriomedial face, the lateral face, or both. The tarsal claws are long and claw tufts very conspicuous.

These are rather small spiders with body lengths usually under 6 mm.

Biology. These spiders run rapidly over plants or are found on the ground. They make tubular silken retreats. *C. abboti* L. Koch is sometimes found in cotton and is listed as a predator on insect eggs (Nyffeler et al. 1990a).

Range. Widespread.

Habitat. These spiders spend the winter under bark and stones.

References. Kaston (1978). Edwards (1958). Levi et al. (1990). Breene et al. (1993b). Platnick (1993).

34 CORINNIDAE—ANTMIMIC SPIDERS

Description

Many of the dull-colored spiders that were in Clubionidae have been transferred to this family. Although typically dull colored, some have a distinct pattern. Several species are ant mimics in appearance and walk like ants.

These spiders resemble Gnaphosidae but can be distinguished by closely spaced, conical front spinnerets. Corinnids have two claws, eyes in two rows, and a prograde arrangement of the legs (Roth 1993).

Biology

Antmimic spiders are nocturnal hunters. No snare is built. They make retreats in litter, under rocks, or in rolled leaves.

Taxonomic Status

Several genera that were formerly in the Clubionidae have now been transferred to other families. The characters that separate these families still seem to be unclear. The genera from Texas that are now considered in Corinnidae are: *Castianeira, Corinna, Mazax, Meriola,* and *Trachelas.*

Outdated and Unofficial Names

Clubionidae (in part).

References

Roth (1993). Reiskind (1969).

34A. *Castianeira*

Description. These spiders resemble large carpenter ants in appearance and behavior. They have been found associated with these ants. They may also mimic velvet ants (mutillid wasps). Usually they move about slowly, like the ants, raising and lowering their abdomens and front legs which simulate the antennae of ants. However, they may run very rapidly when disturbed.

They resemble *Micaria* (now in Gnaphosidae), which also look like ants. They can be separated by the following characters: The thoracic groove is well marked; the first and second tibiae have two or three pairs of ventral spines; both margins of the fang furrows of the chelicerae have two teeth; the labium is wider than long; and the endites lack an oblique depression.

Biology. The eggsacs are flattened discs which are usually attached to the underside of a stone. The outer surface of the eggsac often has a beautiful opaline or pearly luster. They are tough and difficult to tear open.

No snare is built, but retreats are made.

Range. Widespread.

Habitat. Found under stones in pastures, under logs, in wooded areas, and in leaf litter.

Taxonomic Status. About ten species are recorded from Texas, with *amoena* (C. L. Koch), *crocata* (Hentz), *descripta* (Hentz) and *longipalpa* (Hentz) being the most widely spread and commonly reported.

References. Kaston (1978). Reiskind (1969) presents a subfamily work.

34B. *Castianeira descripta* (Hentz), REDSPOTTED ANTMIMIC

Description. The carapace and abdomen are deep mahogany brown to black. The abdomen has red spots often restricted to the posterior end but sometimes extending forward to the anterior end. The legs have dark femora, but the distal segments, especially on the first two legs, are lighter.

Length of the female 8 to 10 mm; length of the male 6.2 to 7.6 mm.

Biology. Like other members of the family, this is a hunting spider that wanders around at night in search of prey.

Range. Reported in east, central, and south Texas. Known from New England and adjacent Canada to Florida west to Texas, Oklahoma, and Iowa.

Habitat. Usually found near the ground.

References. Kaston (1978). Platnick (1993).

34C. *Trachelas*

Description. The sternum and carapace are reddish brown and thickly covered with tiny punctures. The abdomen is pale yellow to light gray, with the anterior median area slightly darker. The legs are darker from front to back.

Biology. Some members of *Trachelas* have been suspected of human envenomation (Uetz 1973, Pase and Jennings 1978).

Range. Widespread.

Habitat. Specimens have been collected from under loose bark of trees and from rolled up leaves, and on many occasions have been found in autumn walking about inside houses. Liao et al. (1984) report *Trachelas* from pecan trees.

Taxonomic Status. All of our species are very similar. Four species are reported from Texas.

References. Kaston (1978). Platnick & Shadab (1974a, 1974b) revised the genus.

35 Gnaphosidae—Ground Spiders

Key Family Characters

Gnaphosids are nocturnal hunting spiders. They are generally uniformly dull colored. Some species have the abdomen patterned with lines or spots. The anterior spinnerets are cylindrical, end abruptly, and are the heaviest and longest spinnerets. The genus *Micaria,* which now is included in this family, has spinnerets more like clubionids.

Gnaphosids have eight eyes, with the posterior median eyes flattened, oval, or irregularly shaped. They have two claws with claw tufts and eight eyes in two rows. The legs have a normal prograde arrangement and are generally spinose.

The cephalothorax has little separation between the cephalic and thoracic parts. The abdomen is elongated or oval and slightly flattened (Roth 1993).

Biology

Gnaphosids spin tubular retreats under rocks, in rolled leaves, or in other sheltered areas. They hide during the day and emerge at night to hunt. They are often captured in pitfall traps.

Eggsacs are thin and papery, white or pink, and flatly attached to the surface at the retreat.

Taxonomic Status

Family revisions include: Heiss and Allen (1986) and Platnick and Shadab (1975, 1976, 1980, 1981, 1982, 1988). *Micaria* has been moved to Gnaphosidae from Clubionidae.

There are 20 genera in Texas.

References

Kaston (1978). Breene et al. (1993b).

Generic Summary for Gnaphosidae in Texas

Quick recognition characters	Genus	Number of species
Cheliceral retromargin with one rounded lobe	*Callilepis*	3
Abdomen with 3–4 longitudinal black lines on pale background	*Cesonia*	2
Deeply notched trochanters	*Drassodes*	2
Embolus long, whip-like; widespread	*Drassyllus*	18
Retromargin of chelicerae with several rounded lobes	*Eilica*	1
Transverse abdominal patterns; cheliceral retromargin with one tooth	*Gertschosa*	1
Cheliceral retromargin with keel	*Gnaphosa*	4
Spinnerets contiguous; abdomen with iridescent lancelolate scales; often ant-like in appearance	*Micaria*	16
Transverse abdominal pattern; cheliceral retromargin lacks denticles and teeth	*Sergiolus*	10
Other genera not easily recognized		
	Haplodrassus	3
	Herpyllus	7
	Nodocion	3
	Scopoides	1
	Scotophaeus	1
	Sosticus	1
	Synaphosus	2
	Talanites	2
	Trachyzelotes	1
	Urozelotes	1
	Zelotes	11

35A. *Drassyllus*

Description. The colors vary from orange to dark brown on the cephalothorax and gray to black on the abdomen. The darker spiders resemble *Zelotes*, with which they have been confused.

Biology. Like other members of the family, these are nocturnal hunting spiders that remain hidden during the day under stones or loose bark.

Range. Most species in this large genus occur in the western states.

Habitat. Found under leaves, stones, and logs on the ground.

Taxonomic Status. The most widespread and frequently recorded species in Texas are: *aprilinus* (Banks), *dromeus* Chamberlin, *lepidus* (Banks), *notonus* Chamberlin, *orgilus* Chamberlin, *prosaphes* Chamberlin, and *texamans* Chamberlin.

References. Kaston (1978). Breene et al. (1993b).

35B. *Gnaphosa*

Description. Most members of this genus have the carapace dark brown with black markings along the radial furrows. The abdomen is dark gray to black and covered with fine hairs. The scutum on the abdomen of the male is small and indistinct.

Biology. *Gnaphosa* are nocturnal hunters. Females are often found with flattened eggsacs containing as many as 250 eggs.

Range. All of our species are quite widespread throughout Texas.

Habitat. Found under stones in pastures and under old logs in wooded areas.

References. Levi et al. (1990). Platnick (1993).

35C. *Herpyllus ecclesiasticus* HENTZ, PARSON SPIDER

Description. The carapace and legs are basically chestnut brown but are covered with indistinct gray markings. The abdomen is brown to black with a light longitudinal stripe down the middle of the front two-thirds of the abdomen. There is also a light spot near the end of the abdomen. The entire body is covered with silky gray hairs.

Length of the female 8 to 13 mm; length of the male 5.5 to 6.5 mm.

Range. Widespread in Texas. New England and adjacent Canada south to Georgia and west to Colorado.

Habitat. This spider can be found under stones and boards, or on the ground. It prefers wooded areas and hibernates under loose bark.

References. Kaston (1978).

35D. *Micaria*

Description. This genus is now placed in Gnaphosidae because the endites have depressions and the posterior median eyes are often oval. The median thoracic groove is lacking or only faintly indicated, and the body is covered with flattened scales, usually brightly colored and iridescent. The color is often lost in preserved material. In many species, the abdomen shows a constriction or depression near its anterior end. The degree of depression varies within a species, and gravid females show less depression.

Biology. *Micaria* are active ground hunters and are noted as ant mimics.

Range. Widespread across the state.

Habitat. These spiders run very rapidly over the ground in dry areas.

Taxonomic Status. This genus is sometimes placed in Clubionidae. The most widespread and frequently recorded species are: *longipes* Emerton, *nanella* Gertsch, *triangulosa* Gertsch, and *vinnula* Gertsch & Davis.

References. Kaston (1978). Levi et al. (1990). Platnick (1989, 1993). Breene et al. (1993b).

35E. *Zelotes*

Description. In general, the colors vary from brownish gray to black without markings. *Zelotes* resemble the darker species of *Drassyllus* and have been confused with them.

Biology. Typical of the family, these are nocturnal hunting spiders.

Range. Widespread across the state.

Habitat. Often found under stones.

Taxonomic Status. The most widespread and frequently recorded species in Texas are: *gertschi* Platnick & Shadab, *hentzi* Barrows, *lasalanus* Chamberlin, and *pseustes* Chamberlin.

References. Kaston (1978). Platnick (1993).

36 CTENIDAE—WANDERING SPIDERS

Key Family Characters

Ctenids are active, day-hunting spiders that resemble the nocturnal wolf spiders. Ctenids have the two eye rows so strongly recurved that the eye arrangement is considered to be 2-4-2, which is unique. The tiny anterior lateral eyes are close to both the posterior median eyes and posterior lateral eyes. The eyes in the back row are larger than the eyes in the front row. In addition, ctenids have two claws with claw tufts and scattered tarsal trichobothria. They are 6 to 13 mm long.

Biology

Eggsacs are reportedly deposited on a substrate or carried by chelicerae. Habits of North American species are almost unknown. Large tropical specimens arrive as hitchhikers in bananas.

Taxonomic Status

The family was studied in North America by Peck (1981). Although found in west, central, south, and southeast Texas, there are relatively few records in the state.

References

Kaston (1978). Levi et al. (1990). Roth (1993).

37 SELENOPIDAE—SELENOPID CRAB SPIDERS

Key Family Characters

Selenopids are large, extremely flat spiders. The eye arrangement is characteristic and consists of six of the eight eyes in a single row in front. These spiders have two tarsal claws with claw tufts, two rows of tarsal trichobothria, and scopulate tarsi and metatarsi.

Biology

Selenopids are nocturnal and move very fast. They can be found on rocky outcrops or in houses in southern states.

Taxonomic Status

Selenopids are primarily a tropical family. There is only one species in Texas, *Selenops actophilus* Chamberlin, which is known from the western part of the state. These spiders are 9 to 12 mm long. A study of the family in North America was completed by Muma (1953).

References

Levi et al. (1990). Roth (1993).

38 HETEROPODIDAE—GIANT CRAB SPIDERS

Key Family Characters

Heteropodids hold two or three legs forward, which gives them their common name. The eyes in the front row are larger than those in the back row. The lateral eyes or all four may be enlarged. These spiders have two claws and eight eyes. These spiders are 10 mm to over 16 mm in size.

Biology

This is a large family of mostly tropical species. They are active nocturnal spiders that hide in crevices in the day and hunt at night. They are often found in and on houses, and are sometimes imported with bananas.

They are not reported as building snares, but webbing is used to form retreats and eggsacs.

Heteropoda females carry eggsacs in their chelicerae. *Olios* females make a tough silken hibernaculum within which they guard the eggsac.

Members of the genus *Olios* have been reported from Texas.

Outdated and Unofficial Names

Sparassidae.

References

Levi et al. (1990). Roth (1993).

38A. *Heteropoda venatoria* (LINNAEUS), HUNTSMAN SPIDER

Description. The carapace is yellow to brown with black pubescence near the hind part. The abdomen is light tan with two or three indistinct longitudinal black lines. The marks are more conspicuous in males.

Length of the female about 23 mm; length of the male about 20 mm.

Biology. This species may occasionally be found in bunches of bananas shipped from Central America, where it is native.

Range. Occurs along the coast in Texas and is also common in Florida.

Habitat. Under bark or in the open on trees. Also found in houses, barns, and other buildings.

Outdated and Unofficial Names. Banana spider.

References. Kaston (1978). Levi et al. (1990). Platnick (1993). Gertsch (1979).

39 PHILODROMIDAE—RUNNING CRAB SPIDERS

Key Family Characters

The bodies of running crab spiders are generally flattened and covered with recumbent setae. The second pair of legs is typically the longest, with the other pairs nearly equal in length. Philodromids have two claws, and eight eyes in two recurved rows. The eyes are not on tubercles. The tarsi have claw tufts and scopulae.

The internal eye structure and features of the developing spiderlings are family characters but difficult to use in the field.

Biology

Philodromids are more abundant in northern states. Running crab spiders are found on foliage and sometimes in cotton.

No snares are built for capturing prey.

Taxonomic Status

The running crab spiders are similar to Thomisidae and considered a subfamily by some authors. There are five genera in North America north of Mexico. Revisions in this family include those by Dondale and Redner (1969, 1978b) and Sauer and Platnick (1972).

References

Kaston (1978). Roth (1993).

39A. *Ebo*

Description. *Ebo* spiders have the second leg at least twice as long as the others. The clypeus is not quite as high as the length of the median ocular area. The anterior median eyes are the largest, with the others subequal, and with the posterior medians closer to the posterior laterals than to each other.

Range. Widespread.

Habitat. Often swept from vegetation.

Taxonomic Status. There are about 20 species in North America. The most widespread species in Texas are: *latithorax* Keyserling; *pepinensis* Gertsch; *punctatus* Sauer and Platnick, and *mexicanus* Banks.

References. Kaston (1978). Breene et al. (1993b).

39B. *Philodromus*

Description. These spiders have very flat bodies which allows them to get under cracks in bark. They are protectively colored and difficult to see except when they move. They often live on plants and can run very rapidly.

Biology. The eggsacs are usually fastened to a leaf, twig, or stone and are guarded by the mother.

Range. Widespread.

Habitat. Usually found on vegetation, including trees.

Taxonomic Status. *Philodromus* is a large and difficult genus to identify to species. There are 17 species in Texas.

References. Kaston (1978).

39C. *Tibellus duttoni* (Hentz)

Description. *T. duttoni* does not resemble a typical philodromid but is highly elongate. The spindly, long, thin legs are often held out fore and aft while at rest. The body is gray or yellowish with a darker longitudinal pattern. There are four spots on the abdomen. The abdomen is 3.5 to 5 times as long as wide. The first tibia has four pairs of ventral spines.

Length of the female 6 to 10 mm; length of the male 5 to 7 mm.

Biology. Members of this species have been found to be predators of cotton fleahoppers (Breene et al. 1989b).

Range. The species occurs in the eastern half of Texas. Reported from New England south to Florida and west to Texas and Minnesota.

Habitat. Usually found on grass.

References. Kaston (1978). Breene et al. (1993b).

THOMISIDAE—CRAB SPIDERS

Key Family Characters

Crab spiders are generally quite flattened in appearance. They hold their legs out to the sides, usually with the front two pairs longer and projecting forward. Crab spiders can walk sideways, which gives them the common name. Some forms are brightly colored and often sit on flowers. Other thomisids are dull or cryptically colored, nocturnal, more secretive, and more likely to be found in association with the bark of trees.

They have eight eyes in two rows which are commonly of different sizes. The lateral eyes are either elevated singly or joined on tubercles. Crab spiders have two claws, with claw tufts usually lacking or sparse. They are covered with sparse, erect setae.

Biology

Crab spiders are typically ambush or sit-and-wait type predators. They capture prey by grabbing and overpowering it. They inject venom to immobilize the prey, and can capture prey that is quite large, sometimes larger than the spider.

Crab spiders are widespread and can be very common. They can be found in nearly any habitat from agricultural fields to woodlots.

Eggsacs are flat and may be attached to a substrate. In most species, females guard the eggsacs but die before the eggs hatch. Young crab spiders feed on soft-bodied insects like flies, small wasps, aphids, and thrips. Larger crab spiders may feed on bees, flies, or even other spiders.

They do not weave snares, retreats, or molting or hibernating nests. The male wraps the female loosely with webbing as part of the courtship ritual.

Taxonomic Status

About 200 species occur in North America. Over 40 species have been reported from Texas.

References

Kaston (1978). Breene et al. (1993b). Levi et al. (1990). Roth (1993).

Generic Summary for Thomisidae in Texas

Quick recognition characters	Genus	Number of species
Brightly colored or lightly colored forms; found on flowers	*Misumena*	1
	Misumenoides	1
	Misumenops	7
	Ozyptila	3
	Synema (in part)	2
Elongate-bodied forms	*Tmarus*	4
Cryptically colored forms	*Xysticus*	19
	Synema (in part)	2
Other genera not easily recognized		
	Bassaniana	3
	Majellula	1

40A. *Misumena vatia* (Clerck), goldenrod crab spider

Description. The female has a white to yellow carapace with the sides somewhat darker than the middle. There is often a tinge of red in the eye region. The abdomen is similar in color to the carapace. It may lack markings or have a bright red band, and sometimes has a median row of spots. The legs are light colored.

The male has a dark reddish brown to red carapace with a creamy white spot in the center extending toward the eyes and clypeus. The first two pairs of legs are reddish brown. The hind two pairs of legs are yellow. The abdomen is creamy white with a pair of dorsal and a pair of lateral red bands.

Length of the female 6 to 9 mm; length of the male 2.9 to 4 mm.

Biology. This species hunts on flowers and can change color to some extent.

Range. Probably occurs throughout Texas, but there are only a few records. Reported to occur in the entire United States and southern Canada.

Habitat. Typically found on flowers, as the common name applies.

Outdated and Unofficial Names. Flower spider, goldenrod spider.

References. Kaston (1978). Platnick (1993).

40B. *Misumenoides formosipes* (Walckenaer)
REDBANDED CRAB SPIDER

Description. The female has a carapace that is creamy white to yellow or yellowish brown, with the sides slightly darker. Some female specimens have broad red bands on the carapace and red spots on the legs. The abdomen may be unmarked, or there may be red or brown markings.

The male has the first two pairs of legs red or brown without lighter rings. The hind two pairs of legs are yellow or white like the abdomen. The coloration is variable, and there may be a background color of white or yellow. Broad red bands may be present on the carapace.

Length of the female 5 to 11.3 mm; length of the male 2.5 to 3.2 mm.

Biology. Eggsacs are from 5 to 10 mm in diameter and are white and lens-shaped, containing 100 or more eggs.

Like the similar *Misumena vatia*, this spider is capable of changing its color to some extent. It is often found near the terminals of plants.

Range. The species has been collected from May through September and is present throughout Texas. Distribution includes the entire United States.

Habitat. This species lives on plants and among flowers. It has been reported from cotton fields.

Outdated and Unofficial Names. *Misumenoides aleatorius.*

References. Kaston (1978). Platnick (1993). Breene et al. (1993b).

40C. *Misumenops asperatus* (Hentz), NORTHERN CRAB SPIDER

Description. The basic color is yellow or white with reddish markings. The dorsal surface is covered with short, rigid hairs which arise from red impressions. The tibia and tarsi of the first leg have red rings.

Length of the female 4.4 to 6 mm; length of the male 3 to 4 mm.

Biology. This brightly colored spider sits on flowers in wait of prey such as insects that alight on the flowers in search of pollen or nectar.

Range. The species occurs in the eastern half of Texas. Reported from the entire United States and southern Canada to Arizona and north to Alberta.

Habitat. In flowers, grass, and foliage.

References. Kaston (1978). Platnick (1993). Breene et al. (1993b).

40D. *Misumenops celer* (Hentz), CELER CRAB SPIDER

Description. The female has a carapace that is white to dull yellow, bright yellow, or even green. The carapace is marked with a white X that extends to the eyes. The legs are light colored in the female. The four front legs are longer than the four rear legs.

The male has the edges of the body marked with red. The first two pairs of legs are ringed with red.

Some abdomens have the back marked with two black or red bands made up of five or six spots. The rear half of the abdomen often has a V-shaped mark pointing toward the rear which is really rows of black or red spots.

Length of the female 5 to 6 mm; length of the male 3 to 4 mm.

Biology. *M. celer* has been considered the most abundant spider in western Texas (Dean and Sterling 1987). Some studies indicate that this species, at times, makes up over half of the spiders in cotton fields.

This species is considered polyphagous (Whitcomb et al. 1963, Muniappan and Chada 1970, Dean et al. 1987). It is considered to be of economic benefit for cotton pest control (Breene et al. 1988a, 1989a, 1990).

Range. Found throughout Texas from May through September.

Habitat. Reported commonly in cotton, corn, and woolly croton.

References. Breene et al. (1993b). Sterling (1982). Knutson and Gilstrap (1989).

40E. *Misumenops oblongus* (Keyserling)

Description. The basic color of the abdomen is light green to whitish or silvery white occasionally margined in red. This species has fewer, less conspicuous spines compared with other members of this genus. The female, in particular, has the carapace with few spines. The carapace is pale greenish to white. Males often show a red marginal stripe on the carapace.

Length of the female 4.9 to 6.2 mm; length of the male 1.5 to 2.6 mm.

Biology. The eggsac has a thin white cover woven over it and contains about 77 eggs. This spider has been collected from May to August.

Range. The species has been reported from most of the state. It is reported as being widespread in the United States but more common in the South.

Habitat. Typically found on vegetation. Reported from cotton fields.

References. Kaston (1978). Breene et al. (1993b).

40F. *Xysticus*

Description. There are numerous species in this genus, and most are various shades of brown or gray with white or yellow markings. Often, there is sexual dimorphism which has created some taxonomic confusion when the sexes were not matched up correctly.

Range. Throughout Texas. Reported from New England south to Georgia and west to the Rockies.

Habitat. These spiders live on and under loose bark, under leaves and stones of the forest floor, and on low plants.

Taxonomic Status. *Xysticus* is a rather large genus, and species determinations are difficult. The most widespread and frequently recorded species in Texas are: *auctificus* Keyserling, *ferox* (Hentz), *funestus* Keyserling, *robinsoni* Gertsch, and *texanus* Banks. Additional information on this genus can be found in Gertsch (1939) and Kaston (1978).

References. Kaston (1978). Breene et al. (1993b).

41 Salticidae—Jumping Spiders

Key Family Characters

Jumping spiders are easily recognized by their eye pattern (Figure 36). They have eight eyes arranged in three (or four) rows. The front row of eyes has the median eyes greatly enlarged. The second and third row of eyes are widespread and moved back on the cephalothorax, nearly forming a square. The front legs are usually the heaviest. Even though the family is easy to determine, there are many genera and species, which complicates further identifications.

Jumping spiders are usually robust and hairy. They may be cryptically colored or boldly marked, even gaudy with bright colors and iridescent scales. Some are exceptionally convincing ant or beetle mimics.

Figure 36. Carapace of a jumping spider (Salticidae) showing eye arrangement.

Biology

Salticids are active during the day and are bold hunters. They are the most visually equipped of all spiders and can be observed stalking their prey. When they are close enough to the prey, they leap onto it and overpower it by force and venom. They can leap several times their own body length. Salticids attach a silk thread to the substrate before they leap after prey or to avoid a predator. If the spider misses its mark, the silk is used as a safety rope for the spider to climb back to the substrate.

Many species constantly tap the surface over which they travel with their pedipalps. These pedipalps probably contain tactile chemoreceptors sensitive to prey semiochemicals (Nyffeler et al. 1990). They feed on a wide range of prey.

Courtship behavior and ornamentation of jumping spiders is elaborate (Forster 1982, Jackson 1982, Richman and Jackson 1992). Males approach females in a zigzag pattern, waving their front legs. Bands on the front legs of males and probably the bright-colored face are important for the female to recognize the male. Males may have a variety of behavior patterns used in courtship, such as tactile displays, pheromones, stridulation, and vibratory signals. After mating, the female constructs an eggsac.

No snares are built, but closely woven retreats are constructed for molting, hibernating, and resting. These retreats are built under bark, between stones, or in rolled leaves. Often, several spiders of the same species build hibernating nests in close proximity. Females place the eggsacs inside silken nests and remain until the spiderlings can disperse.

Taxonomic Status

This is a very large family with 41 genera in Texas. Further information on salticids can be found in Peckham and Peckham (1909), Gertsch (1934), Gertsch and Mulaik (1936), Kaston (1948, 1973, 1978), Barnes (1958), Griswold (1987), and Richman (1989). Richman and Cutler (1978) and Richman (1978) present a checklist and key to the genera of American salticids. Roth (1993) updated the key to genera.

References

Kaston (1978). Levi et al. (1990). Breene et al. (1993b). Roth (1993).

41A. *Corythalia canosa* (Walckenaer)

Description. The carapace is black in the cephalic region and chestnut brown in the thoracic part. There are white scales in a band along each lateral edge and in an elongated spot like an inverted comma behind each rear eye. The dorsum of the abdomen shows a white line on each side in front and two pairs of large black spots behind. There is little difference between the sexes.

Length of the female 5.3 to 5.8 mm; length of the male 5 to 5.2 mm.

Biology. Jackson and MacNab (1989) report a wide variety of display behavior in this species. It feeds on ants.

Range. Central and south Texas. South Carolina south to Florida and west to Texas.

Habitat. In leaf litter and shrubs, and sometimes in homes.

References. Kaston (1978). Jackson and MacNab (1989).

Generic Summary for Salticidae in Texas

Quick recognition characters	Genus	Number of species
Ant-like or pseudoscorpion-like	*Bellota*	2
	Cheliferoides	2
	Paradamoetas	1
	Peckhamia	2
	Sarinda	1
	Synageles	2
	Synemosyna	1
Small; lightly pigmented (carapace may be dark in *Neon*)	*Neon*	1
	Neonella	1
	Talavera	1
Restricted to South Texas (*Bellota* also)	*Bredana*	2
	Rhetenor	1
Abdomen hardened, beetle-like, shiny purple or green	*Agassa*	1
	Sassacus	1
Translucent green; eyes in 4 rows	*Lyssomanes*	1
Bulbous setae on venter of front tibiae	*Thiodina*	2
Largest and hairiest species; iridescent chelicerae; dorsal hair tufts above the anterior median eyes	*Phidippus*	15
A pair of white inverted commas on the cephalothorax	*Corythalia*	1
Elongate or very elongate forms	*Marpissa*	5
Small; zebra stripes or metallic green in males; common around buildings	*Salticus*	2
Large mandibles extending forward (in males); cephalothorax highest at the hind eyes	*Zygoballus*	3

Other genera not easily recognized

Genus	Number of species	Genus	Number of species
Admestina	2	*Plexippus*	1
Bagheera	2	*Poultonella*	2
Eris	5	*Pseudicius*	1
Euophrys	1	*Sitticus*	3
Ghelna	3	*Tutelina*	1
Habrocestum	2	*Tylogonus*	1
Habronattus	18		
Hasarius	1		
Hentzia	2		
Maevia	2		
Menemerus	1		
Messua	1		
Metacyrba	2		
Metaphidippus	3		
Pelegrina	4		
Pellenes	1		
Phlegra	1		
Platycryptus	1		

41B. *Eris*

Description. This genus is intermediate between *Phidippus* and *Metaphidippus* in size and hairiness. The males have the first leg fringed on the underside (Figure 37), and the chelicerae are powerfully developed. The chelicerae often extend forward slightly, and the fangs are sinuate.

Range. Widespread in Texas.

Habitat. Often found on shrubs and trees, especially on terminals.

Taxonomic Status. *Eris militaris* (Hentz) is the most widespread and commonly reported species in the genus in Texas.

Reference. Kaston (1978).

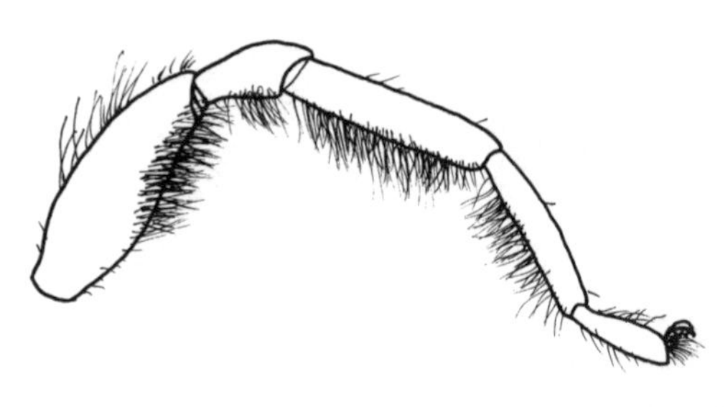

Figure 37. Front leg of male *Eris* showing fringe of hairs on the underside. (*Redrawn from Kaston 1978.*)

41C. *Habronattus*

Description. The male of this genus often has peculiar modifications of form, color, or hairs on the first or third legs or both. These modifications are used for displaying before the female during courtship.

Biology. Biology of this group depends on the species.

Range. Widespread.

Habitat. Found on the ground and upwards on shrubs.

Taxonomic Status. The most widespread species in the state are *texanus* (Chamberlin) and *viridipes* (Hentz). Other widespread and frequently reported species are *coecatus* (Hentz) and *cognatus* (Peckham & Peckham), mostly in East Texas, and *forticulus* (Gertsch & Mulaik) and *mataxus* Griswold, mostly in South Texas.

References. Kaston (1978). Griswold (1987).

41D. *Hentzia palmarum* (HENTZ)

Description. The sexes look quite different in color. The female has gray scales on the dorsal surface and a chevron pattern and oblique rows of small spots at the sides of the dorsal abdomen. The overall appearance is quite elongate for a salticid. Legs are white to yellowish with a translucent appearance.

The male tends to be darker, with two white stripes running from the head region all the way to the spinnerets on either side of the upper surface. The first pair of legs are dark brown, except for the tarsi which are light. The other legs are light like the female. Some males have the chelicerae elongated and projecting forward. These males can be confused with species in the genus *Eris*.

Length of the female 4.7 to 6.1 mm; length of the male 4.0 to 5.3 mm.

Biology. Males, females, and immatures apparently are present throughout the year. Eggsacs are recorded from March to October. Eggs are laid in silken sacs with a clutch size of about 15.

H. palmarum has been noted as a predator of fleahoppers on cotton (Breene et al. 1989b).

Range. The species has been collected from the eastern two-thirds of Texas from June through September. Reported from New England and adjacent Canada south to Florida and west to Oklahoma and Nebraska.

Habitat. Can be found on foliage and is reported from cotton. Usually found on trees, mangroves, scrub oaks, willows, and understory shrubs.

Outdated and Unofficial Names. *Hentzia ambigua* (Walckenaer).

Taxonomic Status. *Hentzia* females have both tufts of hair (called pencils) below the posterior median eyes and spatulate hairs on the ventral retromargin of the patella and distal femur. *H. mitrata* (Hentz) has females with legs unpigmented and is recorded from east Texas.

References. Breene et al. (1993b). Kaston (1978) (as *ambigua*). Richman (1989).

41E. *Lyssomanes viridis* (WALCKENAER), MAGNOLIA GREEN JUMPER

Description. This species is light translucent green throughout the body. The front row of eyes is so strongly recurved that it appears as two separate rows. Eyes in the second, third, and fourth rows are encircled with a black area and some red or yellow scales. The chelicerae of the female are vertical, but those of the male extend forward almost horizontally from the clypeus. Male chelicerae are as long as the carapace.

Length of the female 7 to 8 mm; length of the male 5 to 6 mm.

Biology. This spider has been studied in magnolia trees in Florida (Richman and Whitcomb 1981). In that study, mating took place in May and males disappeared by mid-June. Some females survived until August. Females produced 25–70 eggs laid at heights of 33–131 cm.

Females guard the eggs and only a few produce a second clutch of eggs. The bright green eggs are covered loosely by silk. There are probably seven instars after the postembryonic molts. Males have one less molt than females before maturity. Immatures are present from June until May the following year. Males mature two weeks before females. Prey includes midges, small wasps, small flies, aphids, psocids, and even other salticids (*Hentzia*).

Range. Occurring in eastern Texas, the species is not common on cotton. Reported from North Carolina south to Florida and west to Texas.

Habitat. May be found in low bushes and magnolias.

Taxonomic Status. This species is placed in a separate family, Lyssomanidae, by some authors.

References. Kaston (1978). Levi et al. (1990). Breene et al. (1993b). Richman and Whitcomb (1981).

41F. *Maevia inclemens* (Walckenaer), DIMORPHIC JUMPER

Description. In one variety of male, the body is black and there are three tufts of hairs on the cephalic part. The other variety is more like the female, with red, black, and white markings on a gray background. The female has the abdomen somewhat lighter to yellowish with chevrons on the posterior half.

Length of the female 6.5 to 10 mm; length of the male 4.8 to 7 mm.

Range. Present in east, central, and north central Texas. Reported from New England and adjacent Canada to Florida and west to Texas and Wisconsin.

Habitat. Found in low-growing vegetation.

Outdated and Unofficial Names. *Maevia vittata* (Hentz).

Reference. Kaston (1978).

41G. *Marpissa pikei* (Peckham & Peckham)
Pike slender jumper

Description. *M. pikei* is a narrow, elongate salticid species. The female is light gray or tan with indistinct brown dorsal markings. These markings consist of three thin broken lines or rows of spots. The male has a broad dark band running the length of the abdomen with a more distinct row of black spots.

Length of the female 6.5 to 9.5 mm; length of the male 6 to 8.2 mm.

Range. Found in the eastern half of Texas. Reported from New England to Florida and west to Nebraska, New Mexico, and Arizona.

Habitat. Usually found in grass.

Outdated and Unofficial Names. *Hyctia pikei* Peckham.

References. Breene et al. (1993b). Kaston (1978). Platnick (1989).

41H. *Metacyrba taeniola* (Hentz)

Description. The carapace is black in the ocular area and iridescent mahogany brown on the thoracic part. The posterior declivity of the carapace is limited to the posterior fifth. The abdomen is gray with two rows of whitish yellow narrow spots. The first leg is much thicker than the others and has the femur flattened.

Length of the female 6 to 7 mm; length of the male 5 to 6 mm.

Range. Widely distributed in Texas. Reported from Delaware south to Florida and west to California.

Habitat. Usually found under rocks.

References. Kaston (1978). Platnick (1993).

41I. *Peckhamia picata* (Hentz), antmimic jumper

Description. The general color is reddish brown. It is somewhat darker with violet reflections in the ocular area. The posterior half of the abdomen is shiny black. The carapace has an obvious constriction behind the rear eyes. There is a pair of white spots between the rear eyes.

The abdomen also has a constricted area with another pair of white spots at the sides. The dorsum of the abdomen is covered by a thick, shiny scutum which extends down the sides in both sexes.

Length of the female 3.5 to 5 mm; length of the male 2.8 to 4 mm.

Biology. This species is a remarkable ant mimic.

Range. There are relatively few records scattered across Texas. Reported from New England and adjacent Canada south to Florida and west to Texas and Nebraska.

Habitat. Found on flowers, shrubs, and trees, often in association with ants.

Reference. Kaston (1978).

41J. *Pelegrina galathea* (WALCKENAER), PEPPERED JUMPER

Description. The male has broad white bands stretching from the eyes to the posterior of the dark brown abdomen on either side. He lacks white scales or has only very few on the cymbium and first femur.

The female usually has an indistinct gray and white chevron pattern or pairs of dark spots each preceded by a large white area on the abdomen. The fourth pair of white spots is transverse.

The legs of both sexes are conspicuously ringed.

Length of the female 3.6 to 5.4 mm; length of the male 2.7 to 4.4 mm.

Biology. The species may be quite common at times on cotton and other crops and in pastures and uncultivated areas. *P. galathea* was found to be the second most numerous salticid predator of the cotton fleahopper (Breene et al. 1989a). It also feeds on other small insects and spiders (Horner 1972, Wheeler 1973, Dean et al. 1987).

An average of 158 eggs per eggsac was found by Horner and Starks (1972).

Range. The species has been collected from May through September throughout Texas. It is reported from eastern states and adjacent Canada to the Rockies, but is more common in the south than the north.

Habitat. *Pelegrina* are among the most common jumping spiders and can be collected by sweeping tall grass and bushes.

References. Kaston (1973). Kaston (1978). Platnick (1993). Breene et al. (1993b).

41K. *Phidippus*

Description. The genus *Phidippus* includes our heaviest and hairiest jumping spiders. Males often have "eyebrow" tufts of hairs and most have the chelicerae at least partly iridescent (Figure 38). They usually have a pattern on the abdomen which includes a distinct band, light side bands, and paired white spots above. The size of the spots varies, and those of the second pair are often fused to form a central triangle. The sexes are often colored very differently.

Biology. *Phidippus* are aggressive predators and have been observed pursuing huge prey relative to their size (Gardner 1965). Predation literature may be found in Freed (1984), Roach (1987), and Young (1989a, 1989b).

Range. Widespread.

Habitat. These spiders can be found on the ground, on low vegetation, on shrubs and trees, and indoors. The habitat depends on species and life stage.

References. Kaston (1978). Breene et al. (1993b).

Figure 38. Frontal view of the head of *Phidippus* showing the eye arrangement and hair tufts. *(Redrawn from Kaston 1978.)*

41L. *Phidippus audax* (Hentz), BOLD JUMPER

Description. *P. audax* is a large, mostly black and quite hairy spider. Typically, it has a large white spot on the center of the dorsal abdomen and two smaller posterior spots. These spots can be variable in size and the color may be red, yellow, or orange depending on the age of the spider and local variation. Adults usually have the abdominal spot white. Some specimens have a white band on either side of the carapace. Males have bold white and black rings on the front legs and iridescent green chelicerae.

Length of the female 8 to 15 mm; length of the male 6 to 13 mm.

Biology. This is an extremely common spider and one of those most noticed. It is common in cotton and many other crops (Young and Edwards 1990). *P. audax* has been recorded feeding on boll weevil adults, tarnished plant bugs, and adults and larvae of the bollworm, pink bollworm, tobacco budworm, and cotton leafworm (Kagan 1943, Clark and Glick 1961, Whitcomb et al. 1963, Whitcomb and Bell 1964, Bailey and Chada 1968).

Eggsacs are lenticular, about 9 mm in diameter, and contain from 67 to 218 eggs.

Range. The species is found largely in the eastern half of Texas from May through September. Reported from the Atlantic Coast states and adjacent Canada west to the Rockies and California.

Habitat. Usually found on tree trunks, on vegetation, under stones, and on boards. It often enters houses. The bold behavior and hairy, husky appearance causes concern for homeowners.

Outdated and Unofficial Names. *Phidippus variegatus.*

References. Kaston (1978). Levi et al. (1990). Platnick (1993). Breene et al. (1993b).

41M. *Phidippus cardinalis* (Hentz), CARDINAL JUMPER

Description. *P. cardinalis* is distinguished from most other *Phidippus* by the bright red cephalothorax and abdomen. *P. apacheanus* has the cephalothorax more orange. The thick hairs and bright red give it a velvet-like appearance. A few other species in this genus are also red in color but none are as bright as this species.

Length of female about 9 mm; length of the male about 8 mm.

Biology. This species is much less common than *P. audax* but is very recognizable.

Range. The species has been collected from northern and eastern Texas. Reported from New England south to Florida and west to Texas.

Habitat. Found in low vegetation and sometimes in cotton fields.

References. Kaston (1978). Breene et al. (1993b).

41N. *Phlegra fasciata* (Hahn)

Description. The carapace has two white lines on it. The ocular quadrangle occupies one-third the length of the carapace. The first tibia has three pairs of ventral spines. The abdomen has three white lines separated by broad brown stripes. The male is darker than the female, with brick red hairs in the eye region.

Length of the female 7 to 8 mm; length of the male 6 to 7 mm.

Range. Records in Texas are from central and west Texas. Reported from New England to Florida and west to Texas and Kansas.

Habitat. Found on the surface of leaf litter in scrub areas.

References. Kaston (1978). Platnick (1993).

41O. *Plexippus paykulli* (Audouin), PANTROPICAL JUMPER

Description. The cephalic region of the carapace is black. The thoracic region of the carapace is brown with a light median stripe. The abdomen is black with a yellow median line and white lateral lines. The male has the light median stripe extending farther forward on the carapace and white submarginal lines as well.

Length of the female 10 to 12 mm; length of the male 9.5 mm.

Range. Texas records are from central and south Texas. Reported from Georgia and Florida west to Texas and New Mexico.

Habitat. This species may be found in buildings where it is easily noticed because of its relatively large size and bold markings.

References. Kaston (1978). Levi et al. (1990). Platnick (1993).

41P. *Salticus*

Description. The general color is gray with white markings, but often there are brown to reddish scales mixed in with the gray. Usually, there are iridescent scales in the eye region. Males have more distinct stripes than females which appear more mottled. Males have the chelicerae considerably elongated and extended almost horizontally forward, with the fangs long and sinuate.

Length of the female 4.3 to 6.4 mm; length of the male 4 to 5.5 mm.

Range. *S. scenicus* (Clerck), the zebra jumper, is considered to be found throughout the United States and southern Canada. However, the species recorded in Texas are *austinensis* Gertsch and *peckhamae* (Cockerell). *S. peckhamae* is iridescent green overall and has no white bands.

Habitat. Members of this genus can be very common on fences and outside walls of buildings. They frequently stray indoors.

References. Levi et al. (1990). Platnick (1993). Kaston (1978).

41Q. *Thiodina sylvana* (Hentz)

Description. This species appears quite elongate in form. The front legs are bold and held forward. The carapace is high and rounded, and there are spots of white scales on both sides on the thoracic part. Two white stripes run parallel from the posterior of the carapace to the spinnerets. The two white stripes on the abdomen are bordered by small black spots.

The male has three short white stripes under the posterior lateral eyes, and the abdomen may appear to be dark green.

Length of the female 8 to 10 mm; length of the male 7 to 9 mm.

Range. The species occurs in the eastern half of Texas. Reported from North Carolina south to Florida and west to California. Records from New Mexico, Arizona, and California may be an undescribed species.

Habitat. Usually in trees and shrubs.

Outdated and Unofficial Names. *Thiodina iniquies* (Walckenaer).

References. Kaston (1978). Levi et al. (1990). Breene et al. (1993b).

41R. *Zygoballus rufipes* Peckham & Peckham
HAMMERJAWED JUMPER

Description. The genus can be distinguished from others by the shape of the carapace. It is highest immediately behind the last pair of eyes, then abruptly slopes to the pedicel.

The female has a dark brown carapace. The abdomen is lighter with a white basal transverse band and paired white spots behind. The female has whitish scales that form patterns that are less distinct than the patterns in the male.

The male is darker, with red and green iridescence on the cephalic part of the carapace. The abdomen is bronze-brown with two pairs of white spots parallel to each other. The male has powerfully developed chelicerae with a heavy hammer-like process on the lower surface near the lateral edge.

Length of the female 4.3 to 6 mm; length of the male 3 to 4 mm.

Biology. *Z. rufipes* has been recorded preying upon cotton fleahoppers (Dean et al. 1987).

Range. The species occurs in eastern Texas. Reported from New England to Florida and west to Texas and Nebraska.

Habitat. Members of the genus *Zygoballus* have been noted in moderately high field numbers on cotton and on grasses in some years, while they are nearly absent in other years.

Outdated and Unofficial Names. *Zygoballus bettini* Peckham.

References. Kaston (1978). Platnick (1993). Breene et al. (1993b).

ORDER OPILIONES—HARVESTMEN

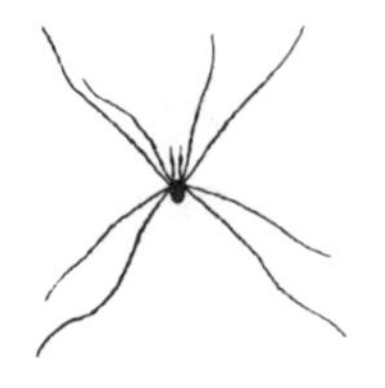

2 HARVESTMEN

Characters

Harvestmen have a globular body. They can be separated from spiders, which have two distinct body segments, because harvestmen have the entire body as one unit. The abdomen is distinctly segmented, and the two eyes are mounted on a large dorsal tubercle on the carapace. While most species have extremely long, spindly legs, there are species with shorter legs.

Biology and Taxonomic Status

Harvestmen are primarily predaceous on insects and other arthropods but sometimes feed on dead insects and plant juices. They have scent glands with ducts to the outside above the first or second coxae. These glands produce a smelly fluid, which may explain the common belief that they are poisonous although no scientific literature verifies that claim.

Some species occur nearly everywhere. They are especially common in wooded areas, under rocks and logs, in caves, and in similar sheltered areas. Worldwide, there are 37 families of harvestmen. Rowland and Reddell (1976) report 18 species of harvestmen in Texas. Members of only one family, Phalangiidae, are properly referred to as daddylonglegs.

References

Rowland and Reddell (1976). Cokendolpher and Lee (1993). Levi et al. (1990).

Order Acari—Ticks

Of the more than 30,000 described species of ticks and mites in the Order Acari, the ticks comprise approximately 800 species. As a group, the Acari occupy habitats as diverse as those of insects and spiders and exhibit every trophic level. This section will focus only on ticks, which are obligate blood-feeding parasites of vertebrates. The direct effects of blood feeding and transmission of disease pathogens make them important in human and animal health.

Characters

The tick body, like spiders, is divided into two regions: the anterior mouthparts (capitulum or gnathosoma) and the general body region that bears the legs and abdomen (idiosoma). However, the body regions are not as distinct as in spiders and have a wide connection, giving ticks a single, more rounded body appearance. Ticks have six legs as larvae and eight legs as nymphs and adults.

Biology and Taxonomic Status

There are two families of ticks in Texas: the Argasidae or "softbacked ticks," and the Ixodidae or "hardbacked ticks." The Argasidae tend to be oval to pear-shaped when viewed from above and are either globular or dorso-ventrally flattened. The integument is unpatterned and has a thick, leathery appearance with a textured surface. The body is marked with folds and grooves that flatten out when they are well fed.

Ixodidae are typically flattened when unfed. The adults have a sclerotized dorsal plate called a scutum which covers the back and is often ornate or patterned. In adult males, the scutum covers the entire dorsal surface behind the capitulum (see Figure 39a). In larvae, nymphs, and adult females, the scutum is restricted to the dorsal area behind the capitulum and over the coxal region (see Figure 39b). The alloscutum is posterior to the scutum in these stages and is less sclerotized, in part enabling considerable expansion during blood feeding.

Both families of ticks have a four-stage life cycle: egg, six-legged larva, eight-legged nymph, and adult. While Ixodidae have only a single nymphal instar, there may be two to seven nymphal instars in Argasidae, depending on the species and various environmental conditions. Ticks feed by making an incision in the skin of a host with their chelicerae. They insert their mouth parts or hypostome. Salivary glands secrete numerous chemical substances which aid in attachment and feeding. All stages of ixodid ticks and some larval and nymphal argasid ticks require long feeding periods (2 to 14 days). Most adult argasids and many immature argasids complete feeding within minutes or a few hours.

Life cycles are categorized by the number of hosts utilized for blood meals or the series of feeding and off-host periods. One-host ticks attach to the host as larvae and complete metamorphosis to nymph and engorged adult on the same host with three blood meals. Two-host ticks attach to the host as larvae, feed, and remain on the host to molt to nymphs and feed again. There are no two-host ticks in North America. Three-host ticks utilize a separate host for each stage–larva, nymph, and adult. Each time a blood meal is completed, the tick drops off the host and molts to the next stage. Adult male ticks generally remain on their host to mate with females and may feed several times. Multi-host ticks, including most soft ticks, have multiple nymphal instars and drop from the host between each instar. These life cycles are somewhat generalized and can be modified to some extent by the conditions.

Ticks are important organisms because their bites are a nuisance, they transmit a number of diseases, and they can be a problem for livestock producers and pet owners.

Reference

Teel (1985).

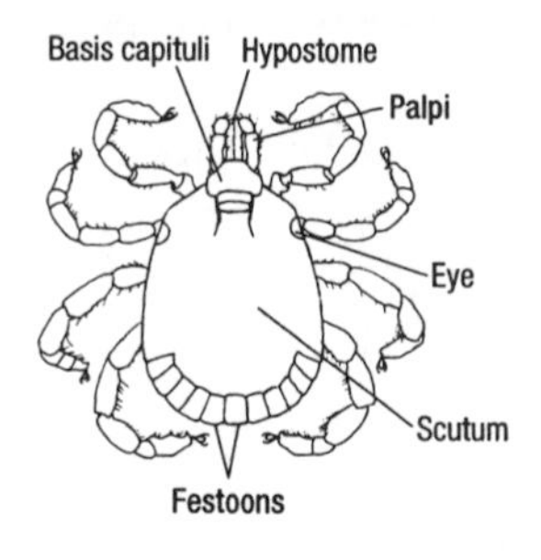

Figure 39a. Dorsal view of male hardbacked tick.

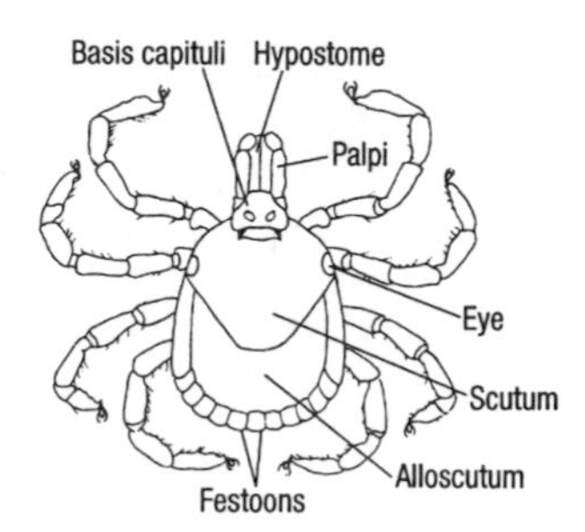

Figure 39b. Dorsal view of female hardbacked tick.

43 IXODIDAE—HARDBACKED TICKS

43A. *Amblyomma americanum* (LINNAEUS), LONE STAR TICK

Description. *A. americanum* are usually ornate, with dark spots and stripes on a pale background. A single white to gold colored spot on the scutum is a good character for field recognition of females. Eyes and festoons are present. These ticks have long mouthparts with palps having unequal segments.

Biology. This tick has a typical three-host life cycle, with probably one generation per year. It overwinters as adult and nymph. Adults are active year round, but exhibit peak activity from March through May. Overwintering nymphs are active between April and June. Nymphal progeny from the adult peak are active from July to October. Larvae have a distinct activity period in July and August. Variation in activity peaks may occur due to geographic and climatic factors. This tick is believed to be a vector of a pathogen causing Lyme disease or a Lyme-like illness and a separate pathogen causing human ehrlichiosis.

Range. Texas through Oklahoma and Missouri and eastward.

Habitat and Hosts. Immature lone star ticks attack rodents, rabbits, ground-dwelling birds, deer, and cattle. Adults as well as immatures will attack deer, cattle, horses, sheep, dogs, and humans.

Lone star ticks are commonly found in wooded areas with underbrush. These habitats provide wildlife hosts and microclimatic conditions for ticks to survive, deposit eggs, and molt.

References. Teel (1985). McDaniel (1979). Hair and Bowman (1986).

43B. *Amblyomma cajennense* (FABRICIUS), CAYENNE TICK

Description. Both males and females are extensively ornate with a diffused pattern. They have long mouth parts with palpal segments that are unequal in size. They have eyes and festoons. Their distribution overlaps that of *Amblyomma imitator*, a tick that closely resembles *A. cajennense*.

Biology. The Cayenne tick has a three-host life cycle. It may be active year round, with nymphs and adults infesting cattle most commonly from March to June. The long mouth parts of this species make their bites quite painful. This tick is a vector of pathogens causing illness similar to Rocky Mountain spotted fever.

Range. Southern part of Texas and south into Mexico and Central and South America.

Habitat and Hosts. Immatures feed on ground-dwelling birds such as quail, coyotes, humans, raccoons, and wild turkey. Adults and immatures attack javelina, deer, cattle, horses, sheep, goats, and dogs.

The off-host ticks are found in brush and drainage areas of rangeland settings.

Reference. Teel (1985).

43C. *Amblyomma maculatum* Koch, Gulf Coast tick

Description. These ticks have eyes, festoons, and long mouth parts with palps having unequal segments. Patterns of scutal ornation are unique.

Biology. The life cycle is that of a typical three-host tick. In coastal and south Texas areas, adult activity begins in the early spring. This activity reaches a peak in August and September. Larvae are active from fall to mid-winter, and nymphs from early winter to spring. In Oklahoma and Kansas, the seasonal pattern of peak activity appears to be about three months earlier.

This species is a nuisance to cattle producers because it tends to feed on the middle and outer parts of cattle ears. Besides the irritation, cattle ears can become thickened and curl inward, a condition known as "gotch ear." As many as 75 to 100 ticks per animal have been observed.

Range. Along the Gulf Coast from South Carolina to Texas and inland into southeastern Kansas extending southward into Mexico and South America.

Habitat and Hosts. Immatures will feed on small mammals, but appear to prefer ground-dwelling birds including quail, meadowlarks, and grasshopper sparrows. Adults feed on cattle, horses, deer, sheep, coyotes, and other carnivores.

Gulf Coast ticks inhabit grasslands and prairies where woody vegetation provides a canopy cover of plants like mesquite, acacia, persimmon, and sumac.

Outdated and Unofficial Names. Sometimes referred to as the "ear tick" by ranchers.

Reference. Teel (1985).

43D. *Dermacentor variabilis* (Say), American dog tick

Description. *Dermacentor* ticks are ornate, with eyes. Mouthparts are short and palpal segments are about equal in size. The basis capituli is rectangular. Scutal patterns of ornation are unique.

Biology. The life cycle is typical of a three-host tick. All stages of the tick may be found on the host year round in moderate climates. In cooler climates, activity patterns are more acute, with adults active in the spring. Immatures are more common from spring to fall.

This tick is an important vector of the Rocky Mountain spotted fever agent and can transmit the pathogen causing anaplasmosis in cattle.

Range. Throughout Texas. Most of the continental United States except the Rocky Mountain region.

Habitat and Hosts. Immatures feed on rodents, particularly the white-footed mouse, meadow mouse, and cotton mouse. Adults prefer dogs and coyotes but will feed on cattle, horses, raccoons, opossums, and humans.

This tick is often encountered in shrub-woods habitats as well as in brush-covered rangeland habitats.

Taxonomic Status. There are a number of other *Dermacentor* tick species in Texas.

References. Teel (1985). McDaniel (1979).

43E. *Ixodes scapularis* Say, black-legged tick

Description. This small tick is recognized in the field by the relatively small size, uniform light body color, and contrasting dark legs. It has long mouthparts with palps having segments of about equal size. No eyes are present.

Biology. The black-legged tick is a three-host tick. Adults are active in the winter and spring. Immatures are active during the spring and summer. This tick is a vector of the pathogen causing Lyme disease.

Range. Most abundant in wooded habitats and drainages of the eastern half of Texas, but encountered less frequently in areas of south central Texas and into Mexico. Extends throughout the southern states, along the east coast to New England and west to the Great Lakes.

Hosts. Immatures feed on a wide variety of small mammals and birds. Adults and immatures feed on deer, cattle, sheep, horses, swine, humans, dogs, cats, coyotes, bobcats, lynx, fox, javelina, rodents, and reptiles.

References. Teel (1985). McIlveen et al. (1990).

43F. *Rhipicephalus sanguineus* (Latreille), brown dog tick

Description. *Rhipicephalus* are inornate with eyes and festoons. They are typically a mottled brown color. Their mouthparts are short, and the basis capituli is hexagonal in shape.

Biology. This tick is well known as a pest of dogs. It can become a serious problem in kennels and urban residential settings. It is a native of Africa but now occurs throughout the world. This tick transmits agents of several diseases to dogs.

Range. Cosmopolitan.

Hosts. This is a three-host tick: dogs are the principal host for all stages, but rodents may also serve as hosts for immature stages. Humans are sometimes attacked when in close association with dogs.

Reference. McDaniel (1979).

44 ARGASIDAE—SOFTBACKED TICKS

44A. *Argas persicus* (OKEN), FOWL TICK

Description. From above, the body is usually oval but sometimes more elliptical when adult. The body margin of these ticks will appear to have a seam joining the dorsal and ventral portions together.

Biology. The life cycle indicates a multihost life cycle. There are two to four nymphal instars that feed once. Adults feed more than once. Fowl ticks tend to be nocturnal and hide in cracks and crevices during the day. These ticks are more commonly found where wood or other structural materials offer harborage off the host.

Fowl tick populations can build up in poultry production situations causing economic loss and even exsanguination of the host. They can transmit several avian pathogens.

Range. Cosmopolitan between 40°N and 40°S latitude.

Hosts. Chickens and other domestic fowl; also wild birds including ducks, geese, and pigeons.

Taxonomic Status. Ticks identified as *A. persicus* may actually be a complex of species.

References. Teel (1985). McDaniel (1979).

44B. *Otobius megnini* (DUGÈS), SPINOSE EAR TICK

Description. Adults have the integument granulated. Nymphs have the integument striated and with spines. No marginal seam is present as in *Argas*.

Biology. This tick is a single-host tick, but the life cycle is somewhat modified. Larvae sit on the ground or on vegetation until a host comes by, then attach and crawl deep into the host's outer ear. They remain there feeding and molt to first and second instar nymphs. The second-instar nymph has a spiny appearance which gives it the common name. The tick may spend from five weeks to several months on the host. The second-instar nymph leaves the host following a blood meal and molts to adulthood on the ground. Adults do not feed, but mate and deposit small batches of eggs on the ground, depending on the previous food reserves. They are frequently found where host animals gather for feeding or resting.

Large numbers or persistent parasitism of immature spinose ear ticks can cause severe irritation or discomfort. Accumulation of matted exuviae and tick feces can result in secondary bacterial infection.

Range. Throughout Texas. From British Columbia through western and southwestern states and into Mexico.

Habitat. May feed on cattle, horses, mules, donkeys, sheep, cats, dogs, wild canines, rabbits, elk, mountain sheep, mountain goats, and even humans.

References. Teel (1985). McDaniel (1979).

Order Scorpiones—Scorpions*

Characters

Scorpions are most easily recognized by the pincers on their pedipalps and their long tails (Figure 40). Scorpions are well equipped to defend themselves or attack prey with their pincers and stingers. In addition to their pincers, they have four pairs of legs. Their bodies are flattened dorso-ventrally, which allows them to hide in small cracks, under rocks, and under bark. Many species dig burrows in the soil.

Scorpions have two eyes on the top of the head and usually two to five pairs at the front corners of the head. Nevertheless, they do not see well, relying on their sense of feel for most activities. Between the last pair of legs is a comblike structure (pectines) that is used to identify substrate textures and for chemoreception (pheromones).

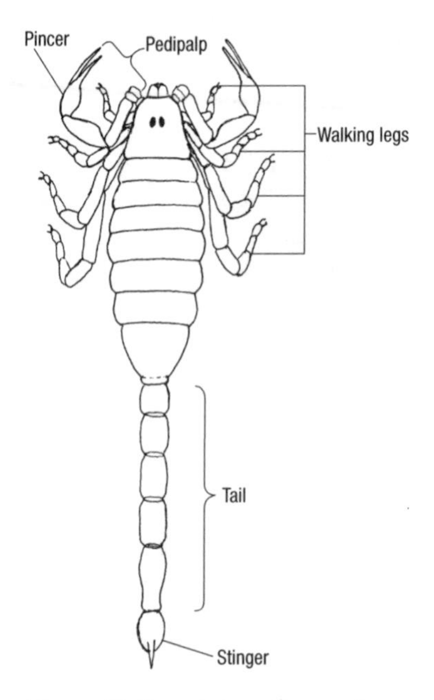

Figure 40. Dorsal view of a scorpion.

*Portions provided by W. D. Sissom.

Biology and Taxonomic Status

Scorpions remain sheltered in the day and become active at night. This behavior helps with thermoregulation and water balance, which are important for their survival in dry habitats. As in other arthropods, their bodies are covered with a waxy cuticle which further helps them reduce moisture loss. For reasons yet unknown, the scorpion cuticle fluoresces under blacklight. Scorpions are capable of reducing their metabolic rates to very low levels and rival spiders in this regard. They store their waste products as uric acid crystals.

Scorpions practice an elaborate courtship in which the partners clasp pincers and dance. Following courtship, the male scorpion deposits a spermatophore from which the female receives a sperm packet. Gestation lasts from a few months to over a year, depending on the species. All scorpions are vivipaous (born live), and embryos are nourished in utero or via a "placental" connection. The average litter size is about 26, which is small compared to many spiders. Female scorpions show considerable parental care. The young climb to the mother's back after birth and soon molt. After the first molt, they disperse to lead independent lives.

Immatures molt an average of six times before maturity. The time to develop to maturity is 6 to 83 months (depending on the species), which is very long for arthropods. Some species may live for 20–25 years, but longevity of the typical scorpion is probably between three to eight years. As adults, they may have several broods.

The sting of scorpions is painful or deadly depending on the species. The venom is a mixture of neurotoxins, which affect the victim's nervous system. Fortunately, none of the species in Texas is considered deadly poisonous. Stings from most of our species are about as painful as a bee or wasp sting, but the severity of the sting is dependent upon the individual scorpion and the person's reaction to the venom. As with any arthropod venom, allergic reactions are possible, and such cases require immediate medical attention. Nonlethal stings may be mild to strong and produce edema, discoloration, numbness, and pain which may last for minutes to several days. Of the 1,500 species of scorpions worldwide, only about 20–25 are regarded as dangerous. These are all in the family Buthidae. Stings from such species may cause paralysis, severe convulsions, cardiac irregularities, and breathing difficulties which may lead to death. Death, if it occurs, usually comes in a matter of hours or perhaps a day after a sting by a lethal species. Antivenins are available in areas where dangerous scorpions occur.

There are about 1,500 species of scorpions in the world, and about 90 known thus far are from the United States. Texas has 18 confirmed species, two of which are undescribed and have not yet been named by scientists.

Only one species, *Centruroides vittatus* (Say), occurs throughout Texas. The number of species increases moving westward and southward in the state, with 1 species recorded in the Dallas area, 2 recorded near Austin, 4 near Amarillo, 3 near Abilene, 5 near Ft. Stockton, 8 in the Ft. Davis region, 8 near Langtry, and 14 in Big Bend National Park.

Scorpions may be found in many types of habitats in the United States, including desert flats, sand dunes, desert and mesic mountains, grasslands, pine forests, deciduous forests, and chaparral. They are most diverse in desert areas.

Key to Families of Scorpions in Texas

Characters	Family
1. Sternum triangular (except in very young specimens); pedipalp patella without ventral trichobothria	Buthidae
2. Sternum pentagonal; pedipalp patella with 2 or 3 ventral trichobothria	
a. Stinger with distinct cone-shaped tubercle under curvature of sting; legs with only one pedal spur at the membrane separating the basitarsus and tarsus	Diplocentridae
b. Stinger without distinct cone-shaped tubercle under curvature of sting; legs with two pedal spurs at the membrane separating the basitarsus and tarsus	Vaejovidae

Species of Scorpions in Texas

Species	Distribution in Texas
Family Buthidae	
Centruroides vittatus (Say)	Statewide
Synonyms: *C. pantheriensis* Stanke	
C. chisosarius Gertsch	
Family Diplocentridae	
Diplocentrus diablo (Stockwell & Nilsson)	Lower Rio Grande Valley
Diplocentrus sp. (undescribed)	Southern Trans-Pecos region
Diplocentrus whitei (Gervais)	Big Bend region
Synonym: *D. bigbendensis* Stahnke	
Family Vaejovidae	
Paruroctonus gracilior (Hoffmann)	Trans-Pecos region
Paruroctonus pecos (Sissom & Francke)	Northern Pecos River drainage
Paruroctonus utahensis (Williams) (erroneously reported as *P. aquilonalis* Stahnke)	Northern part of Trans-Pecos region
Paruroctonus williamsi (Sissom & Francke)	Big Bend National Park
Serradigitus sp. (undescribed)	Big Bend region
Uroctonus apacheanus Gertsch & Soleglad (also known as *Pseudouroctonus apacheanus*)	Trans-Pecos region
Vaejovis chisos Sissom	Chisos Mountains, Big Bend National Park
Vaejovis coahuilae Williams	Most of west Texas
Vaejovis crassimanus Pocock	Trans-Pecos region
Vaejovis globosus Borelli	Along Rio Grande in Big Bend region
Vaejovis intermedius Borelli	Southern parts of Trans-Pecos region to Val Verde County
Vaejovis reddelli Gertsch & Soleglad (also known as *Pseudouroctonus reddelli*)	Texas Hill Country
Vaejovis russelli Williams	Trans-Pecos region and northwest Texas
Vaejovis waueri Gertsch & Soleglad	Western Texas

Note: Some past records of Texas scorpions are based on misidentifications or erroneous locality data. Taxa included in this category are: *Paruroctonus boreus* (Girard), *Vaejovis carolinianus* (Beauvois), *Vaejovis mexicanus* Koch, *Vaejovis bilineatus* Pocock, and *Diplocentrus keyserlingii* Karsch. *Vaejovis spinigerus* (Wood) was presumably described based on specimens originating in Texas, but the species has not been collected in the state since the description was published in 1863.

References: Polis (1990). Sissom (1990). Sissom and Francke (1981, 1985). Shelley and Sissom (1995). Stockwell (1986). Gertsch and Soleglad (1966, 1972).

45 Buthidae—Buthid Scorpions

45A. *Centruroides vittatus* (Say), striped bark scorpion

Description. This species has two broad, blackish longitudinal stripes on the dorsum of the abdomen. Populations in the Big Bend may be only faintly marked or completely pale. The basic color varies from yellowish to tan in adults. Younger specimens may be darker in color and more strongly marked. There is a dark triangular mark on the anterior portion of the carapace in the area over the median and lateral eyes. In young specimens, the bases of the pedipalps and the last segment of the postabdomen are dark brown to black.

The key recognition features are the slender pedipalps and long, slender tail. The tail is longer in males than in females.

Adults average about 60 mm in length.

Biology. These scorpions apparently mate in the fall, spring, and early summer. Gestation estimates are about eight months. Litter size varies from 13–47; the average is about 31.

Immatures molt within three to seven days after birth and remain on the mother for another three to seven days after that. There are five or six molts to maturity. They probably can live for at least four years.

The sting of this species causes local pain and swelling. Deaths attributed to this species are not well substantiated.

Range. This is the most widely distributed scorpion in Texas and the only one in the eastern part of the state.

Habitat. These scorpions are often found under rocks, under boards, and in debris. They can be found indoors or outdoors in a wide variety of habitats (pine forests in East Texas, rocky slopes, grasslands, and juniper breaks in other parts of the state). *Centruroides* are active foragers that do not burrow. They are considered to be "bark scorpions" with a distinct association with dead vegetation, fallen logs, and human dwellings. It is common for them to climb, and many of the reports in homes are associated with attics.

References. Polis (1990). Shelley and Sissom (1995).

46 VAEJOVIDAE—VAEJOVID SCORPIONS

46A. *Paruroctonus gracilior* (HOFFMANN)

Description. This scorpion is yellow brown to greenish brown, usually with a dark triangle on the carapace. The coloration is often obscured because dirt clings readily to the dorsum. The front edge of the çarapace is noticeably convex. The pedipalp pincers are quite robust and heavily keeled in adults, less so in immatures. The tail segments are elongated and the stinger is very slender. The pectines (combs) bear between 23 to 30 teeth in males and 17 to 23 in females.

Adults are between 35 and 45 mm in length.

Biology. Little is known of the biology of this species. It is an obligate burrower. Based on limited observations, the sting is apparently mild.

Range. In Texas, this scorpion is found in the Trans-Pecos region and is fairly common.

Habitat. This species prefers habitats with consolidated sandy or gravelly soils, often in creosote scrub communities. Its burrows are often at the bases of creosote bushes or other shrubs. It can occasionally be found under rocks.

References. Gertsch and Soleglad (1966). Stockwell (1986).

46B. *Paruroctonus utahensis* (WILLIAMS)

Description. This scorpion is very distinctive, being completely pale yellow to light yellow brown (depending on the color of the sand on which it lives). In adults, the pincers are quite swollen with strong keels (ridges) and short fingers. The pectines (combs) bear numerous teeth: 29–37 in males and 17–22 in females. The legs bear distinct bristle combs, which help the scorpion walk and burrow in its loose, sandy habitat.

Biology. Little is known about the general biology of this species, although aspects of its ecology have been studied in New Mexico. It is an obligate burrower, and many important events in its life history, such as birth and molting, occur in the burrow.

Range. *Paruroctonus utahensis* occurs in the northern and extreme western portions of Trans-Pecos Texas. It is abundant in the Monahans Sand Hills, as well as the sand dune systems around El Paso.

Habitat. This species is found only in areas where the substrate consists of loose sands. It is often very abundant in these habitats, where it builds its burrows at the base of vegetation along sand dunes.

References. Sissom and Francke (1981). Stockwell (1986).

46c. *Vaejovis coahuilae* Williams

Description. This species is yellow brown to brown with a distinct dusky pattern on the carapace and dorsum of the abdomen. The underside of the tail bears four dark longitudinal stripes, marking the positions of keels (ridges). The pedipalp pincers are fairly smooth and slightly swollen in males, with short fingers; those of females are of average build, smooth, and with short fingers. The tail is relatively short and stocky.

Adult males average about 30 to 45 mm in length, and females 35 to 50 mm.

Biology. This species is a burrower, but is commonly encountered underneath rocks and other surface debris. Little is known about its reproductive biology, but birth usually occurs from June to August, with litter size ranging from 20 to 41. The young stay on the mother's back for about 9 to 12 days, and maturity probably occurs after the fifth or sixth molt in males and after the sixth or seventh molt in females.

The sting of this species is relatively mild, about as severe as a bee sting or less.

Range. *Vaejovis coahuilae* occurs throughout much of the western half of the state, with isolated records in the Texas Panhandle south to Val Verde County and throughout the Trans-Pecos area. It is quite common, often as common as *Centruroides vittatus.*

Habitat. Although this species is frequently found under rocks, dead yuccas, and other surface objects, it mostly digs spiralling burrows in the soil. These burrows may be located at the bases of shrubs, in open ground, or under rocks. It has been encountered in creosote scrub deserts, grasslands, and rocky slopes over much of its range.

References. Williams (1968). Stockwell (1986).

46D. *Vaejovis reddelli* Gertsch & Soleglad

Description. This scorpion is almost uniformly dark reddish brown to blackish brown, with the stinger often reddish. The pincers are somewhat robust with distinct keels. The tail is relatively slender, and the second through fifth segments are longer than wide. The pectines (combs) bear 16–19 teeth in males and 12–16 teeth in females.

Body size is variable. Specimens found in surface habitats are usually 40–50 mm in length; cave specimens may reach 60 mm in length.

Biology. Little is known about the biology of this species.

Range. *Vaejovis reddelli* is found in south central Texas from Williamson and Burnet counties south to Bexar and Uvalde counties and west to Kimble and Val Verde counties. Much of this range is shared only with *Centruroides vittatus,* rendering its identification quite simple.

Habitat. This species is common in caves throughout the Texas Hill Country. It is also found in surface habitats, particularly under rocks among leaf litter in live oak communities.

Alternate Names. *Pseudouroctonus reddelli* (Gertsch & Soleglad)

References. Gertsch and Soleglad (1972). Stockwell (1986).

46E. *Vaejovis waueri* Gertsch & Soleglad

Description. This small scorpion is yellow brown to brown in color, with a pair of dark stripes down the back and dark pincers. The last tail segment is also darkened. The pincers are small with short fingers, and the tail segments are short and broad, features which easily distinguish this species from small *Centruroides vittatus.* The pectines (combs) bear 12–15 teeth in males, 11–15 in females.

Males are about 18 to 22 mm in length, and females 20 to 25 mm.

Biology. Little is known about the biology of this species. Litter size ranges from 5 to 20, and the young scorpions remain with the mother for about 10 days.

Range. The species occurs in the Trans-Pecos and along the Rio Grande, with sporadic records in northwest Texas. It can be locally common.

Habitat. *Vaejovis waueri* is commonly encountered on rocky or boulder-strewn slopes. It apparently dwells in cracks and crevices among rocks and does not burrow. Specimens can occasionally be found under rocks on slopes.

References. Gertsch and Soleglad (1972). Stockwell (1986).

ORDER UROPYGI—WHIPSCORPIONS

47 THELYPHONIDAE—VINEGAROONS

WHIPSCORPIONS

Key Family Characters

Our Texas species of whipscorpion is a "vinegaroon" in the family Thelyphonidae. Vinegaroons have heavy pedipalps that are formed into pincers. The first pair of legs is long and thin and is used like antennae. The next three pairs of legs are used for walking. The abdomen is attached widely to the cephalothorax. The tail is long and thin, suggesting a whip which is how the common name originates. Our species is nearly black in color.

Adults are 40 to 80 mm or more in body length.

Biology

Vinegaroons can spray a mist from scent glands at the base of the tail when disturbed. The mist contains 85% concentrated acetic acid or vinegar, hence the name. Vinegaroons are nocturnal with poor vision and rely on sensing vibrations to locate prey. They are considered nonpoisonous but can pinch.

Range

The only Texas species is *Mastigoproctus giganteus* (Lucas), the giant vinegaroon. It is primarily found in west Texas, especially in the Trans-Pecos region, but has been reported in south Texas and as far north as the Panhandle.

Habitat

Vinegaroons are usually associated with desert areas but have been reported in grasslands, scrub, pine forests, and mountains.

References

Levi et al. (1990). Rowland and Cooke (1973).

Order Solifugae—Windscorpions

48 Windscorpions

Characters

Windscorpions are from 1 to 5 cm long and most are yellowish to brown in color. They walk on only three pairs of legs. The pedipalps are thin and used like feelers. The first pair of legs are more slender than the others and act as sense organs. The mouth parts of windscorpions are formed into large jaws that work vertically and project forward from the mouth. The appearance of the large, vertically curved jaws is quite distinct.

Biology and Taxonomic Status

Windscorpions are rapidly moving predators that readily attack their prey. They can "run like the wind," hence the name. Although some people erroneously consider them venomous, they do not have venom glands. They are capable of biting.

They may burrow into the sand or hide under stones, and are mostly nocturnal. As in spiders, males are generally smaller and have longer legs. Females bury their eggs, and some guard them. They are short-lived, typically surviving only one year.

Windscorpions are often referred to as Solpugidae by some authors. Rowland and Russell (1976) list 26 species in Texas. There are two families: Ammotrechidae and Eremobatidae. Most of our Texas species are in Eremobatidae, and the largest genus is *Eremobates.* In this family, the front of the head is straight across and the first pair of legs have one or two claws. The species are difficult to identify; many are localized or have records from only a few locations.

Windscorpions are primarily found in deserts and dry areas. They are present throughout the western and southern parts of the state, but are unknown or at least uncommon in east Texas.

Windscorpions have also been referred to as sun spiders, windspiders, and sun scorpions.

References

Levi et al. (1990). Muma (1951, 1962). Rowland and Reddell (1976).

Order Pseudoscorpiones—Pseudoscorpions

49 Pseudoscorpions

PSEUDOSCORPIONS

Characters

The body of the pseudoscorpion is divided into two general parts: the cephalothorax and the abdomen. The body and appendages have many setae. The cephalothorax is covered by a shield or carapace that is not segmented. There is usually one or more pairs of eyes on the lateral edge of the cephalothorax. There are six conspicuous appendages on the cephalothorax: the chelicerae, the palps, and 4 pairs of walking legs.

Chelicerae are short and have a clasping mechanism with a fixed and a movable finger. The pedipalps are longer and have a claw that resembles that of a scorpion. The abdomen has 12 segments, but the last two are reduced and inconspicuous. Pseudoscorpions are quite small, with a body length of generally under 3 mm.

Biology and Taxonomic Status

Pseudoscorpions are predators that feed on a variety of small insects and other arthropods. None are known to be ectoparasites, but they can be found in bird and rodent nests where they feed on arthropods in the nests. They are sometimes found on beetles or large insects where they apparently feed on mites.

They have spinnerets and produce silk that is used to construct nests and temporary sheets for sperm transfer. They do not have a long tail like scorpions and they do not sting.

Pseudoscorpions develop from eggs carried by the female. Eggs hatch into immatures, which are carried by the female. There are three additional immature stages which are free living, and the adult stage. Females usually produce only 3 or 4 eggs, but some produce up to 30 eggs per clutch.

Although data is sparse, adults are apparently long-lived, probably surviving for 6 months to two years.

Pseudoscorpions can be found in a wide variety of habitats, and species preference is well documented. They can be found in ground cover, leaf litter, rotten logs, bark, bogs, swamps, rock outcrops, caves, and homes.

Rowland and Reddell (1976) list 33 species in Texas.

References

Hoff (1949). Weygoldt (1969).

Appendix 1

Checklist of Spiders in Texas*

Mygalomorphae

Atypidae
Sphodros paisano Gertsch & Platnick
Sphodros rufipes (Latreille)

Ctenizidae
Ummidia absoluta (Gertsch & Mulaik)
Ummidia audouini (Lucas)
Ummidia beatula (Gertsch & Mulaik)
Ummidia celsa (Gertsch & Mulaik)
Ummidia funereus (Gertsch)

Cyrtaucheniidae
Eucteniza rex (Chamberlin)
Eucteniza stolida (Gertsch & Mulaik)
Myrmekiaphila comstocki Bishop & Crosby
Myrmekiaphila fluviatilis (Hentz)

Dipluridae
Euagrus chisoseus Gertsch
Euagrus comstocki Gertsch

Theraphosidae
Aphonopelma anax (Chamberlin)
Aphonopelma armada (Chamberlin)
Aphonopelma arnoldi Smith
Aphonopelma breenei Smith
Aphonopelma clarki Smith
Aphonopelma gurleyi Smith
Aphonopelma harlingena (Chamberlin)
Aphonopelma hentzi (Girard)
Aphonopelma heterops (Chamberlin)
Aphonopelma hollyi Smith
Aphonopelma moderatum (Chamberlin & Ivie)
Aphonopelma pseudoroseum (Strand)
Aphonopelma steindacheri (Ausserer)
Aphonopelma texensis (Simon)
Aphonopelma waconum (Chamberlin)

Araneomorphae

Agelenidae
Agelenopsis aleenae Chamberlin & Ivie
Agelenopsis aperta (Gertsch)
Agelenopsis aperta guttata Chamberlin & Ivie
Agelenopsis emertoni Chamberlin & Ivie
Agelenopsis longistylus (Banks)
Agelenopsis naevia (Walckenaer)
Agelenopsis spatula Chamberlin & Ivie
Barronopsis texana (Gertsch)
Tegenaria domestica (Clerck)
Tegenaria pagana C. L. Koch
Tortolena dela Chamberlin & Ivie

Amaurobiidae
Coras alabama Muma
Coras medicinalis (Hentz)
Metaltella simoni (Keyserling)

Anyphaenidae
Anyphaena celer (Hentz)
Anyphaena dixiana (Chamberlin & Woodbury)
Anyphaena fraterna (Banks)
Anyphaena lacka Platnick
Anyphaena pectorosa L. Koch
Hibana arunda (Platnick)
Hibana cambridgei (Bryant)

*Provided by D. A. Dean.

Hibana futilis (Banks)
Hibana gracilis (Hentz)
Hibana incursa (Chamberlin)
Hibana velox (Becker)
Lupettiana mordax (O.P.-Cambridge)
Pippuhana calcar (Bryant)
Wulfila alba (Hentz)
Wulfila bryantae Platnick
Wulfila saltabundus (Hentz)
Wulfila tantillus Chickering

Araneidae
Acacesia hamata (Hentz)
Acanthepeira cherokee Levi
Acanthepeira stellata (Walckenaer)
Agathostichus leucabulba Gertsch
Araneus bicentarius (McCook)
Araneus bonsallae (McCook)
Araneus cavaticus (Keyserling)
Araneus cingulatus (Walckenaer)
Araneus cochise Levi
Araneus detrimentosus (O.P.-Cambridge)
Araneus illaudatus (Gertsch & Mulaik)
Araneus juniperi (Emerton)
Araneus kerr Levi
Araneus marmoreus Clerck
Araneus miniatus (Walckenaer)
Araneus nashoba Levi
Araneus nordmanni (Thorell)
Araneus pegnia (Walckenaer)
Araneus pratensis (Emerton)
Araneus texanus (Archer)
Araniella displicata (Hentz)
Argiope argentata (Fabricius)
Argiope aurantia Lucas
Argiope blanda O. P.-Cambridge
Argiope trifasciata (Forskål)
Colphepeira catawba (Banks)
Cyclosa bifurca (McCook)
Cyclosa caroli (Hentz)
Cyclosa turbinata (Walckenaer)
Cyclosa walckenaeri (O.P.-Cambridge)
Eriophora edax (Blackwall)
Eriophora ravilla (C. L. Koch)
Eustala anastera (Walckenaer)
Eustala bifida F.P.-Cambridge
Eustala brevispina Gertsch & Davis
Eustala cameronensis Gertsch & Davis
Eustala cepina (Walckenaer)
Eustala clavispina (O.P.-Cambridge)
Eustala devia (Gertsch & Mulaik)
Eustala emertoni (Banks)
Gasteracantha cancriformis (Linnaeus)
Gea heptagon (Hentz)
Hypsosinga funebris (Keyserling)
Hypsosinga rubens (Hentz)
Kaira alba (Hentz)
Kaira altiventer O.P.-Cambridge
Kaira hiteae Levi
Larinia directa (Hentz)
Larinioides cornutus (Clerck)
Larinioides patagiatus (Clerck)
Larinioides sclopetarius (Clerck)
Mangora calcarifera F.O.P.-Cambridge
Mangora fascialata Franganillo
Mangora gibberosa (Hentz)
Mangora maculata (Keyserling)
Mangora placida (Hentz)
Mangora spiculata (Hentz)
Mastophora bisaccata Emerton
Mastophora cornigera Hentz
Mastophora phrynosoma Gertsch
Mecynogea lemniscata (Walckenaer)
Metazygia wittfeldae (McCook)
Metazygia zilloides (Banks)
Metepeira arizonica Chamberlin & Ivie
Metepeira comanche Levi
Metepeira foxi Gertsch & Ivie
Metepeira labyrinthea (Hentz)
Metepeira minima Gertsch
Micrathena gracilis (Walckenaer)
Micrathena mitrata (Hentz)
Micrathena sagittata (Walckenaer)
Neoscona arabesca (Walckenaer)
Neoscona crucifera (Lucas)
Neoscona domiciliorum (Hentz)
Neoscona nautica (L. Koch)
Neoscona oaxacensis (Keyserling)
Neoscona utahana (Chamberlin)
Ocrepeira ectypa (Walckenaer)
Ocrepeira georgia (Levi)
Ocrepeira globosa (F.P.-Cambridge)
Ocrepeira redempta (Gertsch & Mulaik)
Scoloderus nigriceps (O. P.-Cambridge)
Verrucosa arenata (Walckenaer)
Wagneriana tauricornis (O.P.-Cambridge)

Caponiidae
Orthonops lapanus Gertsch & Mulaik
Tarsonops systematicus Chamberlin

Clubionidae
Cheiracanthium inclusum (Hentz)
Clubiona abboti L. Koch
Clubiona adjacens Gertsch & Davis

Clubiona catawba Gertsch
Clubiona kagani Gertsch
Clubiona kiowa Gertsch
Clubiona maritima L. Koch
Clubiona pygmaea Banks
Elaver chisosa (Roddy)
Elaver dorothea (Gertsch)
Elaver excepta (L. Koch)
Elaver mulaiki (Gertsch)
Elaver texana (Gertsch)

Corinnidae
Castianeira alteranda Gertsch
Castianeira amoena (C. L. Koch)
Castianeira crocata (Hentz)
Castianeira cubana (Banks)
Castianeira descripta (Hentz)
Castianeira gertschi Kaston
Castianeira longipalpa (Hentz)
Castianeira occidens Reiskind
Castianeira peregrina (Gertsch)
Castianeira trilineata (Hentz)
Corinna sp.
Mazax kaspari Cokendolpher
Mazax pax Reiskind
Meriola deceptus (Banks)
Trachelas mexicanus Banks
Trachelas similis F.O.P.-Cambridge
Trachelas tranquillus (Hentz)
Trachelas volutus Gertsch

Ctenidae
Anahita punctulata (Hentz)
Ctenus valverdiensis Peck
Leptoctenus byrrhus Simon

Dictynidae
Argennina unica Gertsch & Mulaik
Brommella lactea (Chamberlin & Gertsch)
Cicurina aenigma Gertsch
Cicurina arcuata Keyserling
Cicurina armadillo Gertsch
Cicurina bandera Gertsch
Cicurina bandida Gertsch
Cicurina baronia Gertsch
Cicurina barri Gertsch
Cicurina blanco Gertsch
Cicurina browni Gertsch
Cicurina buwata Chamberlin & Ivie
Cicurina caverna Gertsch
Cicurina coryelli Gertsch
Cicurina cueva Gertsch
Cicurina davisi Exline
Cicurina delrio Gertsch
Cicurina dorothea Gertsch
Cicurina elliotti Gertsch
Cicurina ezelli Gertsch
Cicurina gatita Gertsch
Cicurina gruta Gertsch
Cicurina hexops Chamberlin & Ivie
Cicurina holsingeri Gertsch
Cicurina joya Gertsch
Cicurina machete Gertsch
Cicurina madla Gertsch
Cicurina marmorea Gertsch
Cicurina mckenziei Gertsch
Cicurina medina Gertsch
Cicurina menardia Gertsch
Cicurina microps Chamberlin & Ivie
Cicurina minorata (Gertsch & Davis)
Cicurina mirifica Gertsch
Cicurina modesta Gertsch
Cicurina obscura Gertsch
Cicurina orellia Gertsch
Cicurina pablo Gertsch
Cicurina pampa Chamberlin & Ivie
Cicurina pastura Gertsch
Cicurina patei Gertsch
Cicurina porteri Gertsch
Cicurina puentecilla Gertsch
Cicurina rainesi Gertsch
Cicurina reclusa Gertsch
Cicurina reddelli Gertsch
Cicurina reyesi Gertsch
Cicurina riogrande Gertsch & Mulaik
Cicurina robusta Simon
Cicurina rosae Gertsch
Cicurina rudimentops Chamberlin & Ivie
Cicurina russelli Gertsch
Cicurina sansaba Gertsch
Cicurina selecta Gertsch
Cicurina serena Gertsch
Cicurina sheari Gertsch
Cicurina sintonia Gertsch
Cicurina sprousei Gertsch
Cicurina stowersi Gertsch
Cicurina suttoni Gertsch
Cicurina texana (Gertsch)
Cicurina travisae Gertsch
Cicurina ubicki Gertsch
Cicurina uvalde Gertsch
Cicurina varians Gertsch & Mulaik
Cicurina venefica Gertsch
Cicurina venii Gertsch
Cicurina vespera Gertsch
Cicurina vibora Gertsch
Cicurina wartoni Gertsch

Cicurina watersi Gertsch
Dictyna annexa Gertsch & Mulaik
Dictyna bellans Chamberlin
Dictyna bostoniensis Emerton
Dictyna calcarata Banks
Dictyna cholla Gertsch & Davis
Dictyna coloradensis Chamberlin
Dictyna foliacea (Hentz)
Dictyna formidolosa Gertsch & Ivie
Dictyna melva Chamberlin & Gertsch
Dictyna personata Gertsch & Mulaik
Dictyna secuta Chamberlin
Dictyna sylvania Chamberlin & Ivie
Dictyna terrestris Emerton
Dictyna volucripes Keyserling
Emblyna altamira (Gertsch & Davis)
Emblyna callida (Gertsch & Ivie)
Emblyna consulta (Gertsch & Ivie)
Emblyna cruciata (Emerton)
Emblyna evicta (Gertsch & Mulaik)
Emblyna hentzi (Kaston)
Emblyna iviei (Gertsch & Mulaik)
Emblyna orbiculata (Jones)
Emblyna reticulata (Gertsch & Ivie)
Emblyna roscida (Hentz)
Emblyna stulta (Gertsch & Mulaik)
Emblyna sublata (Hentz)
Lathys delicatula (Gertsch & Mulaik)
Lathys maculina Gertsch
Mallos blandus (Chamberlin & Gertsch)
Phantyna bicornis (Emerton)
Phantyna mulegensis (Chamberlin)
Phantyna provida (Gertsch & Mulaik)
Phantyna segregata (Gertsch & Mulaik)
Thallumetus pineus (Chamberlin & Ivie)
Tivyna petrunkevitchi (Gertsch & Mulaik)
Tricholathys knulli Gertsch & Mulaik

Diguetidae
Diguetia albolineata (O.P.-Cambridge)
Diguetia canities (McCook)
Diguetia canities mulaiki Gertsch
Diguetia imperiosa Gertsch & Mulaik

Dysderidae
Dysdera crocota C. Koch

Filistatidae
Filistatinella crassipalpus (Gertsch)
Filistatoides insignis (O.P.-Cambridge)
Kukulcania arizonica (Chamberlin & Ivie)
Kukulcania hibernalis (Hentz)

Gnaphosidae
Callilepis chisos Platnick
Callilepis gertschi Platnick
Callilepis imbecilla (Keyserling)
Cesonia bilineata (Hentz)
Cesonia sincera Gertsch & Mulaik
Drassodes gosiutus Chamberlin
Drassodes saccatus (Emerton)
Drassyllus antonito Platnick & Shadab
Drassyllus aprilinus (Banks)
Drassyllus cerrus Platnick & Shadab
Drassyllus covensis Exline
Drassyllus creolus Chamberlin & Gertsch
Drassyllus dixinus Chamberlin
Drassyllus dromeus Chamberlin
Drassyllus gynosaphes Chamberlin
Drassyllus inanus Chamberlin & Gertsch
Drassyllus lepidus (Banks)
Drassyllus mormon Chamberlin
Drassyllus mumai Gertsch & Riechert
Drassyllus notonus Chamberlin
Drassyllus orgilus Chamberlin
Drassyllus prosaphes Chamberlin
Drassyllus rufulus (Banks)
Drassyllus sinton Platnick & Shadab
Drassyllus texamans Chamberlin
Eilica bicolor Banks
Gertschosa amphiloga (Chamberlin)
Gnaphosa altudona Chamberlin
Gnaphosa clara (Keyserling)
Gnaphosa fontinalis Keyserling
Gnaphosa sericata (L. Koch)
Haplodrassus chamberlini Platnick & Shadab
Haplodrassus dixiensis Chamberlin & Woodbury
Haplodrassus signifer (C. L. Koch)
Herpyllus bubulcus Chamberlin
Herpyllus cockerelli (Banks)
Herpyllus ecclesiasticus Hentz
Herpyllus gertschi Platnick & Shadab
Herpyllus hesperolus Chamberlin
Herpyllus propinquus (Keyserling)
Herpyllus regnans Chamberlin
Micaria deserticola Gertsch
Micaria emertoni Gertsch
Micaria gertschi Barrows & Ivie
Micaria imperiosa Gertsch
Micaria langtry Platnick & Shadab
Micaria longipes Emerton
Micaria mormon Gertsch
Micaria nanella Gertsch

Micaria nye Platnick & Shadab
Micaria palliditarsus Banks
Micaria pasadena Platnick & Shadab
Micaria pulicaria (Sundevall)
Micaria punctata Banks
Micaria seminola Gertsch
Micaria triangulosa Gertsch
Micaria vinnula Gertsch & Davis
Nodocion eclecticus Chamberlin
Nodocion floridanus (Banks)
Nodocion rufithoracicus Worley
Scopoides cambridgei (Gertsch & Davis)
Scotophaeus blackwalli (Thorell)
Sergiolus angustus (Banks)
Sergiolus bicolor Banks
Sergiolus capulatus (Walckenaer)
Sergiolus cyaneiventris Simon
Sergiolus lowelli Chamberlin & Woodbury
Sergiolus minutus (Banks)
Sergiolus montanus (Emerton)
Sergiolus ocellatus (Walckenaer)
Sergiolus stella Chamberlin
Sergiolus tennesseensis Chamberlin
Sosticus insularis (Banks)
Synaphosus paludis (Chamberlin & Gertsch)
Synaphosus syntheticus (Chamberlin)
Talanites captiosus (Gertsch & Davis)
Talanites exlineae (Platnick & Shadab)
Trachyzelotes lyonneti (Audouin)
Urozelotes rusticus (L. Koch)
Zelotes aiken Platnick & Shadab
Zelotes anglo Gertsch & Riechert
Zelotes duplex Chamberlin
Zelotes gertschi Platnick & Shadab
Zelotes hentzi Barrows
Zelotes lasalanus Chamberlin
Zelotes lymnophilus Chamberlin
Zelotes monodens Chamberlin
Zelotes pseustes Chamberlin
Zelotes reformans Chamberlin
Zelotes tuobus Chamberlin

Hahniidae
Hahnia arizonica Chamberlin & Ivie
Hahnia cinerea Emerton
Hahnia flaviceps Emerton
Neoantistea agilis (Keyserling)
Neoantistea mulaiki Gertsch
Neoantistea oklahomensis Opell & Beatty

Hersiliidae
Tama mexicana (O.P.-Cambridge)

Heteropodidae
Heteropoda sp.
Olios sp.

Leptonetidae
Archoleptoneta garza Gertsch
Neoleptoneta anopica (Gertsch)
Neoleptoneta chisosea (Gertsch)
Neoleptoneta coeca (Chamberlin & Ivie)
Neoleptoneta concinna (Gertsch)
Neoleptoneta devia (Gertsch)
Neoleptoneta furtiva (Gertsch)
Neoleptoneta microps (Gertsch)
Neoleptoneta myopica (Gertsch)
Neoleptoneta uvaldea (Gertsch)
Neoleptoneta valverdae (Gertsch)

Linyphiidae
Centromerus latidens (Emerton)
Ceraticelus emertoni (O.P.-Cambridge)
Ceraticelus paludigenus Crosby & Bishop
Ceraticelus paschalis Crosby & Bishop
Ceratinops crenatus (Emerton)
Ceratinops rugosus (Emerton)
Ceratinopsis laticeps Emerton
Ceratinopsis purpurescens Keyserling
Eperigone antrea (Crosby)
Eperigone bryantae Ivie & Barrows
Eperigone eschatologica (Crosby)
Eperigone maculata (Banks)
Eperigone paula Millidge
Eperigone tridentata (Emerton)
Erigone autumnalis Emerton
Erigone barrowsi Crosby & Bishop
Erigone dentigera O.P.-Cambridge
Erigone personata Gertsch & Davis
Eulaira suspecta Gertsch & Mulaik
Floricomus mulaiki Gertsch & Davis
Floricomus ornatulus Gertsch & Ivie
Floricomus rostratus (Emerton)
Florinda coccinea (Hentz)
Frontinella communis Hentz
Grammonota inornata Emerton
Grammonota maculata Banks
Grammonota nigrifrons Gertsch & Mulaik
Grammonota suspiciosa Gertsch & Mulaik
Grammonota texana (Banks)
Grammonota vittata Barrows

Idionella anomala (Gertsch & Ivie)
Idionella deserta (Gertsch & Ivie)
Idionella formosa (Banks)
Idionella sclerata (Ivie & Barrows)
Islandiana flaveola (Banks)
Islandiana unicornis Ivie
Lepthyphantes sabulosus (Keyserling)
Lepthyphantes zebra (Emerton)
Masoncus conspectus (Gertsch & Davis)
Meioneta fabra (Keyserling)
Meioneta llanoensis (Gertsch & Davis)
Meioneta micaria (Emerton)
Neriene radiata (Walckenaer)
Soulgas corticarius (Emerton)
Tapinocyba hortensis (Emerton)
Tennesseellum formica (Emerton)
Tutaibo anglicanus (Hentz)
Walckenaeria puella Millidge
Walckenaeria spiralis (Emerton)

Liocranidae

Phrurolithus formica Banks
Phrurotimpus alarius (Hentz)
Phrurotimpus apertus (Gertsch)
Phrurotimpus borealis (Emerton)
Phrurotimpus callidus (Gertsch)
Phrurotimpus emertoni (Gertsch)
Phrurotimpus leviculus (Gertsch)
Scotinella fratrellus (Gertsch)

Lycosidae

Allocosa absoluta (Gertsch)
Allocosa apora (Gertsch)
Allocosa funerea (Hentz)
Allocosa furtiva (Gertsch)
Allocosa mulaiki (Gertsch)
Allocosa noctuabunda (Montgomery)
Allocosa pylora Chamberlin
Alopecosa kochi (Keyserling)
Arctosa littoralis (Hentz)
Arctosa minuta F.O.P-Cambridge
Geolycosa fatifera (Hentz)
Geolycosa latifrons Montgomery
Geolycosa missouriensis (Banks)
Geolycosa riograndae Wallace
Geolycosa sepulchralis (Montgomery)
Gladicosa euepigynata (Montgomery)
Gladicosa gulosa (Walckenaer)
Gladicosa pulchra (Keyserling)
Hesperocosa unica (Gertsch & Wallace)
Hogna annexa (Chamberlin & Ivie)
Hogna antelucana (Montgomery)
Hogna aspersa (Hentz)
Hogna benedicta (Chamberlin)
Hogna carolinensis (Walckenaer)
Hogna coloradensis (Banks)
Hogna georgicola (Walckenaer)
Hogna helluo (Walckenaer)
Hogna retenta (Gertsch & Wallace)
Isohogna lenta (Hentz)
Isohogna tigana (Gertsch & Wallace)
Pardosa atlantica Emerton
Pardosa delicatula Gertsch & Wallace
Pardosa falcifera F.O.P.-Cambridge
Pardosa littoralis Banks
Pardosa mercurialis Montgomery
Pardosa milvina (Hentz)
Pardosa pauxilla Montgomery
Pardosa sierra Banks
Pardosa sternalis (Thorell)
Pardosa vadosa Barnes
Pardosa zionis Chamberlin & Ivie
Pirata davisi Wallace & Exline
Pirata felix O.P.-Cambridge
Pirata hiteorum Wallace & Exline
Pirata piraticus (Clerck)
Pirata sedentarius Montgomery
Pirata seminolus Gertsch & Wallace
Rabidosa punctulata (Hentz)
Rabidosa rabida (Walckenaer)
Schizocosa aulonia Dondale
Schizocosa avida (Walckenaer)
Schizocosa bilineata (Emerton)
Schizocosa mccooki (Montgomery)
Schizocosa ocreata (Hentz)
Schizocosa retrorsa (Banks)
Schizocosa saltatrix (Hentz)
Schizocosa segregata Gertsch & Wallace
Sosippus texanus Brady
Trochosa gosiuta (Chamberlin)
Trochosa parthenus (Chamberlin)
Trochosa terricola Thorell
Varacosa acompa (Chamberlin)
Varacosa avara (Keyserling)
Varacosa shenandoa (Chamberlin & Ivie)

Mimetidae

Ero canionus Chamberlin & Ivie
Ero pensacolae Ivie & Barrows
Mimetus haynesi Gertsch & Mulaik
Mimetus hesperus Chamberlin
Mimetus notius Chamberlin
Mimetus puritanus Chamberlin
Mimetus syllepsicus Hentz

Miturgidae

Strotarchus piscatorius (Hentz)
Strotarchus planeticus Edwards
Syspira longipes (Simon)
Teminius affinis Banks

Mysmenidae

Calodipoena incredula Gertsch & Davis

Nesticidae

Eidmannella bullata Gertsch
Eidmannella delicata Gertsch
Eidmannella nasuta Gertsch
Eidmannella pallida (Emerton)
Eidmannella reclusa Gertsch
Eidmannella rostrata Gertsch
Gaucelmus augustinus Keyserling

Oecobiidae

Oecobius annulipes Lucas
Oecobius cellariorum (Dugès)
Oecobius putus O.P.-Cambridge

Oonopidae

Oonops furtivus Gertsch
Oonops secretus Gertsch
Oonops stylifer Gertsch
Opopaea devia Gertsch
Opopaea meditata Gertsch & Davis
Opopaea sedata Gertsch & Mulaik
Orchestina saltitans Banks
Scaphiella hespera Chamberlin
Scaphiella juvenilis (Gertsch & Davis)

Oxyopidae

Hamataliwa grisea Keyserling
Hamataliwa helia (Chamberlin)
Hamataliwa unca Brady
Oxyopes acleistus Chamberlin
Oxyopes aglossus Chamberlin
Oxyopes apollo Brady
Oxyopes lynx Brady
Oxyopes salticus Hentz
Oxyopes scalaris Hentz
Oxyopes tridens Brady
Peucetia longipalpis F.O.P.-Cambridge
Peucetia viridans (Hentz)

Philodromidae

Apollophanes punctipes (O.P.-Cambridge)
Apollophanes texanus Banks
Ebo albocaudatus Schick
Ebo evansae Sauer & Platnick
Ebo latithorax Keyserling
Ebo merkeli Schick
Ebo mexicanus Banks
Ebo pepinensis Gertsch
Ebo punctatus Sauer & Platnick
Ebo redneri Cokendolpher
Ebo texanus (Gertsch)
Philodromus alascensis Keyserling
Philodromus cespitum (Walckenaer)
Philodromus histrio (Latreille)
Philodromus imbecillus Keyserling
Philodromus infuscatus infuscatus Keyserling
Philodromus keyserlingi Marx
Philodromus laticeps Keyserling
Philodromus marginellus Banks
Philodromus marxi Keyserling
Philodromus minutus Banks
Philodromus montanus Bryant
Philodromus placidus Banks
Philodromus praelustris Keyserling
Philodromus pratariae (Scheffer)
Philodromus rufus quartus Dondale & Redner
Philodromus undarum Barnes
Philodromus vulgaris (Hentz)
Thanatus formicinus (Clerck)
Thanatus rubicellus Mello-Leitão
Thanatus vulgaris Simon
Tibellus duttoni (Hentz)
Tibellus oblongus (Walckenaer)

Pholcidae

Crossopriza stridulans Millot
Metagonia caudata O.P.-Cambridge
Micropholcus fauroti (Simon)
Modisimus texanus Banks
Pholcophora diluta Gertsch & Mulaik
Pholcophora texana Gertsch
Pholcus phalangioides (Fuesslin)
Physocyclus enaulus Crosby
Physocyclus globosus (Taczanowski)
Physocyclus hoogstraali Gertsch & Davis
Psilochorus coahuilanus Gertsch & Davis
Psilochorus imitatus Gertsch & Mulaik
Psilochorus pallidulus Gertsch
Psilochorus pullulus (Hentz)
Psilochorus redemptus Gertsch & Mulaik
Psilochorus utahensis Chamberlin
Smeringopus pallidus (Blackwall)
Spermophora senoculata (Dugès)

Pisauridae

Dolomedes albineus Hentz
Dolomedes scriptus Hentz
Dolomedes tenebrosus Hentz
Dolomedes triton (Walckenaer)
Dolomedes vittatus Walckenaer
Pisaurina dubia (Hentz)
Pisaurina mira (Walckenaer)
Tinus peregrinus (Bishop)

Prodidomidae

Prodidomus rufus Hentz

Salticidae

Admestina archboldi Piel
Admestina tibialis (C. L. Koch)
Agassa cyanea (Hentz)
Bagheera felix (Peckham & Peckham)
Bagheera prosper (Peckham)
Bellota micans Peckham & Peckham
Bellota wheeleri Peckham & Peckham
Bredana alternata Gertsch
Bredana complicata Gertsch
Cheliferoides longimanus Gertsch
Cheliferoides segmentatus F.O.P.-Cambridge
Corythalia canosa (Walckenaer)
Eris aurantia (Lucas)
Eris fartilis (Peckham & Peckham)
Eris flava (Peckham & Peckham)
Eris militaris (Hentz)
Eris pinea (Kaston)
Euophrys diminuta (Banks)
Ghelna barrowsi (Kaston)
Ghelna castanea (Hentz)
Ghelna sexmaculata (Banks)
Habrocestum acerbum Peckham & Peckham
Habrocestum pulex (Hentz)
Habronattus calcaratus agricola Griswold
Habronattus coecatus (Hentz)
Habronattus cognatus (Peckham & Peckham)
Habronattus delectus (Peckham & Peckham)
Habronattus dorotheae (Gertsch & Mulaik)
Habronattus fallax (Peckham & Peckham)
Habronattus forticulus (Gertsch & Mulaik)
Habronattus hirsutus (Peckham & Peckham)
Habronattus mataxus Griswold
Habronattus mexicanus (Peckham & Peckham)
Habronattus moratus (Gertsch & Mulaik)
Habronattus orbus Griswold
Habronattus sugillatus Griswold
Habronattus texanus (Chamberlin)
Habronattus tranquillus (Peckham & Peckham)
Habronattus tuberculatus (Gertsch & Mulaik)
Habronattus virgulatus Griswold
Habronattus viridipes (Hentz)
Hasarius adansoni (Audouin)
Hentzia mitrata (Hentz)
Hentzia palmarum (Hentz)
Lyssomanes viridis (Walckenaer)
Maevia inclemens (Walckenaer)
Maevia poultoni Peckham
Marpissa bryantae (Jones)
Marpissa formosa (Banks)
Marpissa lineata (C. L. Koch)
Marpissa obtusa Barnes
Marpissa pikei (Peckham & Peckham)
Menemerus bivittatus (Dufour)
Messua limbata (Banks)
Metacyrba punctata (Peckham & Peckham)
Metacyrba taeniola (Hentz)
Metaphidippus chera Chamberlin
Metaphidippus texanus (Banks)
Metaphidippus vitis (Cockerell)
Neon nellii Peckham & Peckham
Neonella vinnula Gertsch
Paradamoetas formicina Peckham & Peckham
Peckhamia americana (Peckham & Peckham)
Peckhamia picata (Hentz)
Pelegrina arizonensis (Peckham & Peckham)
Pelegrina chalceola Maddison
Pelegrina exiguus (Banks)
Pelegrina galathea (Walckenaer)
Pelegrina peckhamorum (Kaston)
Pelegrina pervaga (Peckham & Peckham)
Pelegrina proterva (Walckenaer)
Pelegrina sabinema Maddison
Pelegrina tillandsiae (Kaston)
Pellenes limatus Peckham & Peckham

Phidippus apacheanus Chamberlin & Gertsch
Phidippus arizonensis (Peckham & Peckham)
Phidippus audax (Hentz)
Phidippus cardinalis (Hentz)
Phidippus carneus Peckham
Phidippus carolinensis Peckham & Peckham
Phidippus clarus Keyserling
Phidippus comatus Peckham & Peckham
Phidippus mystaceus (Hentz)
Phidippus otiosus (Hentz)
Phidippus pius Scheffer
Phidippus princeps (Peckham & Peckham)
Phidippus pruinosus Peckham & Peckham
Phidippus putnami (Peckham & Peckham)
Phidippus texanus Banks
Phlegra fasciata (Hahn)
Platycryptus undatus (DeGeer)
Plexippus paykulli (Audouin)
Poultonella alboimmaculata (Peckham & Peckham)
Poultonella nuecesensis Cokendolpher & Horner
Pseudicius piraticus (Peckham & Peckham)
Rhetenor texanus Gertsch
Salticus austinensis Gertsch
Salticus peckhamae (Cockerell)
Sarinda hentzi (Banks)
Sassacus papenhoei Peckham & Peckham
Sitticus cf. *cursor* Barrows
Sitticus dorsatus (Banks)
Sitticus welchi Gertsch & Mulaik
Synageles bishopi Cutler
Synageles noxiosus (Hentz)
Synemosyna formica (Hentz)
Talavera minuta (Banks)
Thiodina puerpera (Hentz)
Thiodina sylvana (Hentz)
Tutelina elegans (Hentz)
Tylogonus minutus (F.O.P.-Cambridge)
Zygoballus nervosus (Peckham & Peckham)
Zygoballus rufipes Peckham & Peckham
Zygoballus sexpunctatus (Hentz)

Scytodidae
Scytodes championi F.O.P-Cambridge
Scytodes dorothea Gertsch
Scytodes perfecta Banks
Scytodes thoracica (Latreille)
Scytodes zapatana Gertsch & Mulaik

Segestriidae
Ariadna bicolor (Hentz)

Selenopidae
Selenops actophilus Chamberlin

Sicariidae
Loxosceles apachea Gertsch & Ennik
Loxosceles blanda Gertsch & Ennik
Loxosceles devia Gertsch & Mulaik
Loxosceles reclusa Gertsch & Mulaik
Loxosceles rufescens (Dufour)

Tengellidae
Zorocrates aemulus Gertsch
Zorocrates alternatus Gertsch & Davis
Zorocrates isolatus Gertsch & Davis

Tetragnathidae
Azilia affinis O.P.-Cambridge
Glenognatha foxi (McCook)
Leucauge venusta (Walckenaer)
Meta mimetoides (Chamberlin & Ivie)
Nephila clavipes (Linnaeus)
Pachygnatha tristriata C.L. Koch
Tetragnatha caudata Emerton
Tetragnatha elongata Walckenaer
Tetragnatha guatemalensis O.P.-Cambridge
Tetragnatha laboriosa Hentz
Tetragnatha nitens (Audouin)
Tetragnatha pallescens F.O.P.-Cambridge
Tetragnatha straminea Emerton
Tetragnatha vermiformis Emerton
Tetragnatha versicolor Walckenaer
Tetragnatha viridis Walckenaer

Theridiidae
Achaearanea florendida Levi
Achaearanea globosa (Hentz)
Achaearanea insulsa (Gertsch & Mulaik)
Achaearanea porteri (Banks)
Achaearanea schullei (Gertsch & Mulaik)
Achaearanea tepidariorum (C. L. Koch)

Anelosimus studiosus (Hentz)
Argyrodes americanus (Taczanowski)
Argyrodes cancellatus (Hentz)
Argyrodes caudatus (Taczanowski)
Argyrodes davisi Exline & Levi
Argyrodes elevatus Taczanowski
Argyrodes fictilium (Hentz)
Argyrodes furcatus (O.P.-Cambridge)
Argyrodes globosus Keyserling
Argyrodes pluto Banks
Argyrodes projiciens (O.P.-Cambridge)
Argyrodes subdolus O.P.-Cambridge
Argyrodes trigonum (Hentz)
Chrosiothes jocosus (Gertsch & Davis)
Chrosiothes minusculus (Gertsch)
Chrysso albomaculata O.P.-Cambridge
Coleosoma acutiventer (Keyserling)
Coleosoma adamsoni (Berland)
Crustulina altera Gertsch & Archer
Crustulina sticta (O.P.-Cambridge)
Dipoena abdita Gertsch & Mulaik
Dipoena alta Keyserling
Dipoena cathedralis Levi
Dipoena nigra (Emerton)
Enoplognatha marmorata (Hentz)
Enoplognatha tecta (Keyserling)
Episinus cognatus O.P.-Cambridge
Euryopis lineatipes O.P.-Cambridge
Euryopis mulaiki Levi
Euryopis quinquemaculata Banks
Euryopis spinigera O.P.-Cambridge
Euryopis taczanowskii Keyserling
Euryopis texana Banks
Latrodectus geometricus C. L. Koch
Latrodectus hesperus Chamberlin & Ivie
Latrodectus mactans (Fabricius)
Latrodectus variolus Walckenaer
Nesticoides rufipes (Lucas)
Phoroncidia americana (Emerton)
Spintharus flavidus Hentz
Steatoda americana (Emerton)
Steatoda borealis (Hentz)
Steatoda fulva (Keyserling)
Steatoda medialis (Banks)
Steatoda mexicana Levi
Steatoda pulcher (Keyserling)
Steatoda quadrimaculata (O.P.-Cambridge)
Steatoda transversa (Banks)
Steatoda triangulosa (Walckenaer)
Stemmops bicolor O.P.-Cambridge
Theridion alabamense Gertsch & Archer
Theridion antonii Keyserling
Theridion australe Banks
Theridion cameronense Levi
Theridion cinctipes Banks
Theridion cynicum Gertsch & Mulaik
Theridion differens Emerton
Theridion dilutum Levi
Theridion dividuum Gertsch & Archer
Theridion flavonotatum Becker
Theridion glaucescens Becker
Theridion goodnightorum Levi
Theridion hidalgo Levi
Theridion llano Levi
Theridion lyricum Walckenaer
Theridion murarium Emerton
Theridion myersi Levi
Theridion positivum Chamberlin
Theridion punctosparsum Emerton
Theridion rabuni Chamberlin & Ivie
Theridion rufipes Lucas
Theridion submissum Gertsch & Davis
Theridula opulenta (Walckenaer)
Thymoites expulsus (Gertsch & Mulaik)
Thymoites illudens (Gertsch & Mulaik)
Thymoites marxi (Crosby)
Thymoites missionensis (Levi)
Thymoites pallidus (Emerton)
Thymoites unimaculatus (Emerton)
Tidarren haemorrhoidale (Bertkau)
Tidarren sisyphoides (Walckenaer)
Wamba crispula (Simon)

Thomisidae
Bassaniana floridana (Banks)
Bassaniana utahensis (Gertsch)
Bassaniana versicolor (Keyserling)
Majellula sp.
Misumena vatia (Clerck)
Misumenoides formosipes (Walckenaer)
Misumenops asperatus (Hentz)
Misumenops californicus (Banks)
Misumenops carletonicus Dondale & Redner
Misumenops celer (Hentz)
Misumenops coloradensis Gertsch
Misumenops dubius Keyserling
Misumenops oblongus (Keyserling)
Ozyptila americana Banks
Ozyptila hardyi Gertsch
Ozyptila monroensis Keyserling
Synema parvulum (Hentz)
Synema viridans (Banks)

Tmarus angulatus (Walckenaer)
Tmarus floridensis Keyserling
Tmarus rubromaculatus Keyserling
Tmarus unicus Gertsch
Xysticus apachecus Gertsch
Xysticus aprilinus Bryant
Xysticus auctificus Keyserling
Xysticus concursus Gertsch
Xysticus elegans Keyserling
Xysticus ellipticus Turnbull, Dondale, & Redner
Xysticus emertoni Keyserling
Xysticus ferox (Hentz)
Xysticus fraternus Banks
Xysticus funestus Keyserling
Xysticus furtivus Gertsch
Xysticus gulosus Keyserling
Xysticus lassanus Chamberlin
Xysticus locuples Keyserling
Xysticus nevadensis (Keyserling)
Xysticus pellax O.P.-Cambridge
Xysticus punctatus Keyserling
Xysticus robinsoni Gertsch
Xysticus texanus Banks

Titanoecidae
Titanoeca americana Emerton
Titanoeca nigrella (Chamberlin)

Uloboridae
Hyptiotes cavatus (Hentz)
Hyptiotes puebla Muma & Gertsch
Miagrammopes mexicanus O.P.-Cambridge
Philoponella oweni (Chamberlin)
Philoponella semiplumosa (Simon)
Uloborus glomosus (Walckenaer)
Uloborus segregatus Gertsch

Zoridae
Zora pumila (Hentz)

Appendix 2

Threatened and Endangered Species of Arachnids in Texas

Scientific Name	Common Name*	Global, Federal, State Status
Archeolarca guadalupensis	Guadalupe Cave Pseudoscorpion	G1, C2, S1
Cicurina bandida	Bandit Cave Spider	G1, C2, S1
Cicurina baronia	Robber Baron Cave Spider	G1, C2, S1
Cicurina cueva	A Cave Spider	G1, C2, S1
Cicurina madla	Madla's Cave Spider	G1, C2, S1
Cicurina venii	Veni's Cave Spider	G1, C2, S1
Cicurina vespera	Vesper Cave Spider	G1, C2, S1
Cicurina wartoni	Warton's Cave Spider	G1, C1, S1
Neoleptoneta microps	Government Canyon Cave Spider	G1, C2, S1
Neoleptoneta myopica	Tooth Cave Spider	G1, LE, S1
Tartarocreagris texana	Tooth Cave Pseudoscorpion	G1, LE, S1
Texella cokendolpheri	Robber Baron Cave Harvestman	G1, C2, S1
Texella reddelli	Bee Creek Cave Harvestman	G1, LE, S1
Texella reyesi	Bone Cave Harvestman	G1Q, LE, S1

*These common names are not recognized by the AmericanTarantula Society.

Global Rank

G1–Critically imperiled globally, extremely rare, 5 or fewer occurrences. (Critically endangered throughout range.)

Q–Qualifier denoting questionable taxonomic assignment.

Federal Rank

LE–Listed Endangered.

C1–Candidate, Category 1. USFWS has substantial information on biological vulnerability and threats to support proposing to list as endangered or threatened. Data are being gathered on habitat needs and/or critical habitat designations.

C2–Candidate, Category 2. Information indicates that proposing to list as endangered or threatened is possibly appropriate, but substantial data on biological vulnerability and threats are not currently known to support the immediate preparation of rules. Further biological research and field study will be necessary to ascertain the status and/or taxonomic validity of the taxa in Category 2.

State Rank

S1–Critically imperiled in state, extremely rare, very vulnerable to extirpation, 5 or fewer occurrences.

Source: Texas Parks and Wildlife Dept., Texas Biological and Conservation Data System, Special Animal List, Jan. 29, 1996.

APPENDIX 3

AMERICAN TARANTULA SOCIETY

This society was begun in 1978 and continued until 1984. The society produced the newsletter *Tarantula Times*. The newsletter, as well as the society, had difficulty staying active. There were 21 issues of the newsletter published from 1978 to 1984. The purpose of the American Tarantula Society was to provide the opportunity for professionals and laymen to share their knowledge of the tarantula, to encourage the study of the tarantula as it gains popularity as a pet, and to eliminate misunderstanding concerning the tarantula. A later goal of the society was to have the "pet industry" breed their own supply of tarantulas, eliminating the plunder of natural habitats and populations. Although the society is no longer active, the problem of wild tarantula collecting for resale continues.

A second American Tarantula Society was founded in 1991 to promote the study and dissemination of information concerning the infraorder Mygalomorphae, especially, but not limited to, the family Theraphosidae, and to maintain a flow of information and cooperation between enthusiasts and professional arachnologists world-wide. The society publishes *Forum of the American Tarantula Society* (ISSN 1062-9718) six times a year. The *Forum* is written in an entertaining style easily understood by the hobbyist and contains articles, news items, and advertisements for the sale and exchange of materials, including live stock. For further information, contact: Membership Secretary, American Tarantula Society, P.O. Box 756, Carlsbad, NM 88221-0756, E-mail: rgbreene@aol.com. Additional information can also be found on the ATS home page at: http://torgo.cnchost.com/ats/.

Glossary*

Abdomen–the second, or posterior, portion of the body of the spider.

Accessory claws–the serrated bristles near the true claws on the tarsi of some spiders.

ALE–anterior lateral eyes.

AME–anterior median eyes.

Annulate–showing rings of pigmentation, as of a leg.

Anteapical–positioned just before the apex or tip.

Anterior–toward the front.

Apodeme–the body wall invagination serving as a muscle attachment area.

Apophysis–an evagination, more stout than a spine, typically on the legs or pedipalps.

Appendages–structures extending away from the body proper, as legs, palps, etc.

Arachnida–class in the phylum Arthropoda that includes spiders, harvestmen, ticks and mites, scorpions, whipscorpions, windscorpions, and pseudoscorpions.

Arachnology–the scientific study of arachnids.

Araneae–the arachnid order of spiders.

Araneologist–a biologist who specializes in the study of spiders.

Araneomorphae (Araneae)–one of the two infraorders of spiders (the other is Mygalomorphae).

Araneophagy–predation upon spiders.

Autotomy–the process whereby a spider breaks off a leg being held by an enemy.

Basitarsus–same as metatarsus.

Book lung–a respiratory organ with page-like folds, found in most spiders.

Boss–a smooth prominence generally at the lateral angle of the base of chelicera, found in some spiders.

Bristle–a long, thin extension of the cuticle, more slender than a spine.

Calamistrum–a series of curved bristles on the dorsal surface or retrolateral edge of metatarsus 4, of some (cribellate) spiders.

Carapace–the fused series of sclerites making up the dorsal part of the cephalothorax.

---

*Modified from Kaston 1978 and Breene et al. 1993b.

Carina–a keel, as on the clypeus or the chelicerae, in some spiders.

Caudad–positioned toward the tail; posterior.

Caudal–a tail or posterior end.

Cephalothorax–the anterior of the two major divisions into which the body of a spider is divided.

Cervical groove–the furrow which extends forward and toward the sides from the center of the carapace and marks the boundary between the head and the thorax; it is sometimes indistinct or completely lacking.

Chela–a pincer-like appendage as typified by scorpions; also, a pincer-like arrangement of the fang with the lamella on the basal segment of the chelicerae of spiders.

Chelicera–the front jaws, consisting of a stout basal segment and a terminal fang.

Chitin–a nitrogenous polysaccharide occurring in the cuticle of arthropods.

Claw–a strong curved process at the distal end of the leg or palp.

Claw tufts–the bunch of hairs at the tip of the tarsus in those spiders with only two claws.

Clypeus–the space between the anterior row of eyes and the anterior edge of the carapace.

Colulus–an appendage resembling a spinneret positioned anterior of the spinnerets; not a non-silk-spinning appendage.

Comb–single bristles with barbs that make up a comb on tarsus 4 in theridiids and nesticids; used to "comb out" silk onto prey.

Conspecific–members of the same species.

Coxa–the segment of the leg (or pedipalp) nearest the body.

Cribellate–adjective referring to a spider that possesses a cribellum.

Cribellum–a silk-spinning, transverse, plate-like organ in front of the spinnerets in cribellate spiders; from it issues the so-called hackled band threads.

Cursorial–adapted for walking or running.

Cymbium–the tarsus of the male pedipalp hollowed out to contain the copulatory organ.

Declivity–a decline, as that which occurs at the rear of the carapace.

Denticle–a small, smooth tooth, usually on chelicerae, legs, or palps.

Diad–a pair, as of two eyes placed close together.

Dorsal–situated near the top or above other sections.

Dorsum–the back or upper surface.

Ecribellate–adjective referring to a spider that does not possess a cribellum.

Edaphic–of or relating to the soil.

Eggsac–spider eggs enclosed in silk.

Embolus–the portion of the male copulatory organ through which the sperm is passed into the seminal receptacle of the female.

Endite–one of the mouth parts, ventral to the mouth opening and lateral to the lip, so that in chewing it opposes the chelicerae.

Entomophagous–feeding on insects.

Epigastric furrow–a groove separating the region of the book lung in the labidognath spider from the more posterior portion of the venter.

Epigynum–a ventral abdominal sclerite of the female reproductive openings.

Exuviae–the cast "skin," i. e., the old exoskeleton of an arthropod.

Fangs–claw-like segments of the spider chelicerae.

Femur–the third segment of the pedipalp or leg, counting from the proximal end of these appendages.

Folium–pigmented design or pattern on the dorsal abdomen, often shaped like a leaf.

Generalist predator–a predator that may attack many different types of prey.

Geniculate–elbowed or bent at a right angle.

Guild–all taxa in a community that use similar resources such as food or space.

Heterogeneous–the characteristic wherein some eyes (usually the AME) are dark in color; the remaining eyes are light in color.

Homogeneous–the condition in which all eyes are the same color.

Immature–a nonadult arthropod.

Instar–the stage of the arthropod between successive molts, e. g., the fourth instar.

Intraguild–existing among different species of a guild.

Kleptoparasitic–the stealing of prey caught by another predator.

Labium–the lower lip between the two endites of spiders.

Lamella–a triangular plate on the promargin of the cheliceral fang furrow in some spiders. It resembles a broad tooth and forms a kind of chela with the fang.

Lamelliform–flattened, as of certain hairs in claw tufts.

Lamina–a flat plate.

Laterigrade–a sideways type of locomotion, as in the crab spiders and their allies: also, the way the legs are turned in on these spiders so that the morphological dorsal surface is posterior.

Lorum–a set of plates on the dorsal side of the pedicel.

Mastidion–a small denticle or tubercle on the anterior face of the chelicera in certain spiders.

Median ocular area–the space limited by the four median eyes.

Metatarsus–the sixth segment of the leg, counting from nearest the body.

Oophagy–predation upon eggs.

Palp–the segments of the pedipalp distal to the endite or coxa. In females it resembles a leg; in males it is modified for sperm transfer.

Papillae–tubercle extensions.

Paracymbium–an accessory branch of the cymbium arising from the proximal part of the latter in some spiders.

Patella–the fourth segment of the leg or pedipalp, counting from the proximal end.

Pedicel–the small stalk connecting the abdomen to the cephalothorax.

Pedipalp–the second appendage of the cephalothorax, behind the chelicerae but in front of the legs.

Phytophagous–feeding on plant materials.

PLE–posterior lateral eyes.

PME–posterior median eyes.

Procurved–a curved arc, typically of an eye row, such that the ends are nearer than its center to the front of the body (see recurved).

Promargin–the margin of the cheliceral fang furrow closer to the front of the body, away from the endite (see retromargin).

Raptorial–adapted for grasping prey with the front legs.

Rebordered–with a thickened edge, as of the labium of some spiders.

Recurved–a curved arc such that the ends are nearer than its center to the posterior of the body (see procurved).

Retromargin–the margin of the cheliceral fang furrow farther from the front of the body, nearer the endite (see promargin).

Saltatorial–adapted for jumping.

Scape–an appendage, free at one end (usually the posterior) and lying in the midline of the epigynum.

Sclerite–a hardened body wall plate bounded by sutures or membranes.

Scopula–a brush of hairs on the lower surface of the tarsus and metatarsus in some spiders.

Scutum–a sclerotized plate, as on the abdomen of some spiders.

Serrated bristle–a type of bristle, usually curved slightly, bearing teeth along one side and forming accessory claws in some spiders.

Sigillum–an impressed area on the sternum.

Spermatheca–a sperm storage organ in females.

Spinnerets–the silk-spinning, paired appendages on the end of the abdomen.

Spinose–provided with spines.

Spiracle–the opening of the tubular tracheae on the ventral side of the abdomen.

Spur–a cuticular process, heavier than a spine.

Spurious claws–the serrated bristles at the end of the tarsus.

Stabilimentum–the bands of silk spun by certain orbweaver species in their webs.

Sternum–the central plate on the underside of the cephalothorax of a spider.

Stridulating organ–an area with numerous parallel setae, which is rubbed by an opposing structure on the pedipalp or abdomen, thus producing a sound.

Sub–a prefix indicating "almost," as in subterminal, subglobose, subequal, subequidistant, subadult, etc.

Tarsus–the last segment of the leg or pedipalp.

Thorax–that portion of the cephalothorax posterior to the cervical groove.

Tibia–the fifth segment of the leg or pedipalp, counting from the proximal end.

Tracheae–tubes through which air is carried around the body of the spider and which open at the spiracle (or spiracles).

Triad–a group of three, as of three eyes placed together.

Trichobothrium–a fine sensory hair protruding at a right angle from a leg.

Trochanter–the second segment of the leg or pedipalp, counting from the proximal end.

Truncate–referring to the end of a part being squared off, instead of coming to a point.

Tubercle–a low, usually rounded, process.

Univoltine–having a single generation per year.

Venter–the underside.

Bibliography

Agnew, C. W., D. A. Dean, and J. W. Smith, Jr. 1985. Spiders collected from peanuts and non-agricultural habitats in the Texas west cross-timbers. *Southwest. Nat.* 30: 1–12.

Agnew, C. W., and J. W. Smith, Jr. 1989. Ecology of spiders (Araneae) in a peanut agroecosystem. *Environ. Entomol.* 18: 30–42.

Bailey, C. L., and H. L. Chada. 1968. Spider populations in grain sorghums. *Ann. Entomol. Soc. Am.* 61: 567–571.

Barnes, R. D. 1958. North American jumping spiders of the subfamily Marpissinae (Araneae, Salticidae). *Am. Mus. Novit.* 1867: 1–50.

Beatty, J. A. 1970. The spider genus *Ariadna* in the Americas (Araneae, Dysderidae). *Bull. Mus. Comp. Zool.* 139: 433–517.

Berman, J. D., and H. W. Levi. 1971. The orb weaver genus *Neoscona* in North America (Araneae: Araneidae). *Bull. Mus. Comp. Zool.* 141: 465–500.

Brach, V. 1977. *Anelosimus studiosus* (Araneae: Theridiidae) and the evolution of quasisociality in theridiid spiders. *Evolution* 31: 154–161.

Brady, A. R. 1964. The lynx spiders of North America, north of Mexico (Araneae: Oxyopidae). *Bull. Mus. Comp. Zool.* 131: 429–518.

Breene, R. G., III. 1988. Predation ecology and the natural control of *Pseudatomoscelis seriatus,* (Hemiptera: Miridae). Ph. D. dissertation, Texas A&M University, College Station.

Breene, R. G., D. A. Dean, G. B. Edwards, B. Hebert, H. W. Levi, G. Manning, and L. Sorkin. 1995. *Common Names of Arachnids.* The American Tarantula Society, 94 pp.

Breene, R. G., D. A. Dean, and R. L. Meagher, Jr. 1993a. Spiders and ants of Texas citrus groves. *Fla. Entomol.* 76: 168–170.

Breene, R. G., D. A. Dean, M. Nyffeler, and G. B. Edwards. 1993b. Biology, predation ecology, and significance of spiders in Texas cotton ecosystems with a key to the species. Texas Agric. Exp. St., College Station. B-1711.

Breene, R. G., R. L. Meagher, Jr., and D. A. Dean. 1993c. Spiders (Araneae) and ants (Hymenoptera: Formicidae) in Texas sugarcane fields. *Fla. Entomol.* 76: 645–650.

Breene, R. G., and W. L. Sterling. 1988. Quantitative phosphorus-32 labeling method for analysis of predators of the cotton fleahopper (Hemiptera: Miridae). *J. Econ. Entomol.* 81: 1494–1498.

Breene, R. G., and M. H. Sweet. 1985. Evidence of insemination of multiple females by the male black widow spider *Latrodectus mactans* (Araneae: Theridiidae). *J. Arachnol.* 13: 331–336.

Breene, R. G., W. L. Sterling, and D. A. Dean. 1988a. Spider and ant predators of the cotton fleahopper on woolly croton. *Southwest. Entomol.* 13: 177–183.

Breene, R. C., M. H. Sweet, and J. K. Olson. 1988b. Spider predators of mosquito larvae. *J. Arachnol.* 16: 275–277.

Breene, R. G., A. W. Hartstack, W. L. Sterling, and M. Nyffeler. 1989a. Natural control of the cotton fleahopper (Hemiptera: Miridae) in Texas. *J. Appl. Entomol.* 108: 298–305.

Breene, R. G., W. L. Sterling, and D. A. Dean. 1989b. Predators of the cotton fleahopper on cotton (Hemiptera: Miridae). *Southwest. Entomol.* 14: 159–166.

Breene, R. G., W. L. Sterling, and M. Nyffeler. 1990. Efficacy of spider and ant predators on the cotton fleahopper [Hemiptera: Miridae]. *Entomophaga* 35: 393–401.

Brignoli, P. M. 1977. Ragni del Brasile III. Note su *Bruchonops melloi.* Biraben e sulla posizione sistematica dei Caponiidae (Arachnida, Araneae). *Revue Suisse Zool.* 84: 609–616.

Brignoli, P. M. 1983. *A Catalogue of the Araneae Described Between 1940 and 1981.* Manchester Univ. Press, Manchester, England, 755 pp.

Brown, K. M. 1974. A preliminary checklist of spiders of Nacogdoches, Texas. *J. Arachnol.* 1: 229–240.

Bruce, J. A., and J. E. Carico. 1988. Silk use during mating in *Pisaurina mira* Walckenaer (Araneae: Pisauridae). *J. Arachnol.* 16: 1–4.

Bumroongsook, S., M. K. Harris, and D. A. Dean. 1992. Predation on blackmargined aphids (Homoptera: Aphididae) by spiders on pecan. *Biol. Control* 2:15–18.

Buskirk, K. E. 1981. Sociality in the Arachnida. In: *Social Insects, Vol. 11.* Academic Press, London, pp. 281– 367.

Carico, J. E. 1973. The nearctic species of the genus *Dolomedes* (Araneae: Pisauridae). *Bull. Mus. Comp. Zool.* 144: 435–488.

Carpenter, R. M. 1972. The jumping spiders (Salticidae) of Wichita County, Texas. *Southwest. Nat.* 17: 161–168.

Chamberlin, R. V. 1923. The North American species of *Mimetus. J. Entomol. Zool.* 15: 3–9.

Chamberlin, R. V., and W. J. Gertsch. 1958. The spider family Dictynidae in America north of Mexico. *Bull. Am. Mus. Nat. Hist.* 116: 1–152.

Chamberlin, R. V., and R. L. Hoffman. 1958. Checklist of the millipedes of North America. U. S. National Museum, Bull. 212, 236 pp.

Chickering, A. M. 1969. Oonopidae of Florida. *Psyche* 76: 144–162.

Clark, E. W., and P. A. Glick. 1961. Some predators and scavengers feeding upon pink bollworm moths. *J. Econ. Entomol.* 54: 815–816.

Coddington, J. A., and H. W. Levi. 1991. Systematics and evolution of spiders (Araneae). *Annu. Rev. Ecol. Syst.* 22: 565–592.

Cokendolpher, J. C., N. V. Horner, and D. T. Jennings. 1979. Crab spiders of north-central Texas (Araneae: Philodromidae and Thomisidae). *J. Kansas Entomol. Soc.* 52: 723–734.

Cokendolpher, J. C., and V. F. Lee. 1993. *Catalogue of the Cyphopalpatores and Bibliography of the Harvestmen* (Arachnida, Opiliones) of Greenland, Canada, U. S. A., and Mexico. Vintage Press, Lubbock, Texas. 82 pp.

Comstock, J. H. 1940. *The Spider Book.* (ed. by W. J. Gertsch). Comstock Publ. Assoc., Ithaca, New York, 729 pp., reprinted 1975.

Corey, D. T., and D. J. Mott. 1991. A revision of the genus *Zoras* (Araneae, Zoridae) in North America. *J. Arachnol.* 19: 55–61.

Coyle, F. A. 1988. A revision of the American funnel-web mygalomorph spider genus *Euagrus* (Araneae, Dipluridae). *Bull. Am. Mus. Nat. Hist.* 187: 203–292.

Craig, C. L., and G. D. Bernard. 1990. Insect attraction to ultraviolet-reflecting spider webs and web decorations. *Ecology* 71: 616–623.

Culin, J. D., and K. V. Yeargan. 1982. Feeding behaviour and prey of *Neoscona arabesca* (Araneae: Araneidae) and *Tetragnatha laboriosa* (Araneae: Tetragnathidae) in soybean fields. *Entomophaga* 27: 417–424.

Dean, D. A., and W. L. Sterling. 1987. Distribution and abundance patterns of spiders inhabiting cotton in Texas. Texas Agric. Exp. Stn. Bull. 1566, College Station.

Dean, D. A., and W. L. Sterling. 1990. Seasonal patterns of spiders captured in a suction trap. *Southwest. Entomol.* 15: 399–412.

Dean, D. A., W. L. Sterling, and N. V. Horner. 1982. Spiders in eastern Texas cotton fields. *J. Arachnol.* 10: 251–260.

Dean, D. A., W. L. Sterling, M. Nyffeler, and R. G. Breene. 1987. Foraging by selected spider predators on the cotton fleahopper and other prey. *Southwest. Entomol.* 12: 263–270.

Dean, D. A., M. Nyffeler, and W. L. Sterling. 1988. Natural enemies of spiders: mud dauber wasps (Hymenoptera: Sphecidae) in east Texas. *Southwest. Entomol.* 13: 283–290.

Deevey, G. B. 1949. The developmental history of *Latrodectus mactans* (Fabr.) at different rates of feeding. *Am. Midl. Nat.* 42: 189–219.

Demmler, G. J., M. L. Levy, C. L. Cole, C. O. Mishaw, A. B. Benson, R. M. Thaller, and M. Feingold. 1989. Picture of the month. *AJDC.* 143: 843–844.

Dondale, C. D., and J. H. Redner. 1969. The *infuscatus* and *dispar* groups of the spider genus *Philodromus* in North and Central America and the West Indies (Araneida: Thomisidae). *Can. Entomol.* 101: 921–954.

Dondale, C. D., and J. H. Redner. 1978a. Revision of the nearctic wolf spider genus *Schizocosa* (Araneida: Lycosidae). *Can. Entomol.* 110: 143–181.

Dondale, C. D., and J. H. Redner. 1978b. The insects and arachnids of Canada–part 5. The crab spiders of Canada and Alaska (Araneae: Philodromidae and Thomisidae). Publ. Dept. Agric. Can. 1663:1–255.

Dondale, C. D., and J. H. Redner. 1983. The wolf spider genus *Allocosa* in North and Central America (Araneae: Lycosidae). *Can. Entomol.* 115: 933–964.

Dondale, C. D., and J. H. Redner. 1984. Revision of the *milvina* group of the wolf spider genus *Pardosa* (Araneae: Lycosidae). *Psyche* 91: 67–117.

Edwards, G. B. 1983. The southern house spider, *Filistata hibernalis* Hentz, (Araneae: Filistatidae). Florida Dept. Agric. Cons. Serv. DPI Entomol. Circ. 255: 1–2.

Edwards, G. B. 1986. A tropical orbweaver, *Eriophora ravilla* (Araneae: Araneidae). Florida Dept. Agric. Cons. Serv. DPI Entomol. Circ. 286: 1–2.

Edwards, R. J. 1958. The spider subfamily Clubioninae of the United States, Canada, and Alaska (Araneae: Clubionidae). *Bull. Mus. Comp. Zool.* 118: 365–436.

Eisner, T., R. Alsop, and P. Ettershank. 1964. Adhesiveness of spider silk. *Science* (Washington, DC) 146: 1058–1061.

Exline, H., and H. W. Levi. 1962. American spiders of the genus *Argyrodes* (Araneae, Theridiidae). *Bull. Mus. Comp. Zool.* 127: 75–204.

Exline, H., and W. H. Whitcomb. 1965. Clarification of the mating procedure of *Peucetia viridans* (Araneida: Oxyopidae) by a microscopic examination of the epigynal plug. *Florida Entomol.* 48: 169–171.

Fink, L. S. 1984. Venom spitting by the green lynx spider, *Peucetia viridans* (Araneae, Oxyopidae). *J. Arachnol.* 12: 372–373.

Fink, L. S. 1986. Costs and benefits of maternal behavior in the green lynx spider (Oxyopidae, *Peucetia viridans*). *Anim. Behav.* 34: 1051–1060.

Fitch, H. S. 1963. Spiders of the University of Kansas natural history reservation and Rockefeller Experimental Tract. Univ. of Kansas Mus. of Nat. Hist. Misc. Publ . 33, Lawrence, 202 pp.

Forster, L. M. 1982. Vision and prey-catching strategies in jumping spiders. *Am. Scientist* 70: 165–175.

Forster, L. M. 1992. The interplay behaviour of sexual cannibalism in *Latrodectus hasselti* (Araneae, Theridiidae), the Australian redback spider. *Austr. J. Zool.* 40: 1 –11.

Forster, R. R., and N. I. Platnick. 1985. A review of the austral family Orsolobidae (Arachnida, Araneae) with notes on the superfamily Dysderoidea. *Bull. Am. Mus. Nat. Hist.* 181: 1–229.

Freed, A. N. 1984. Foraging behaviour in the jumping spider *Phidippus audax:* bases for selectivity. *J. Zool.* (London) 203: 49–62.

Gardner, B. T. 1965. Observations on three species of *Phidippus* jumping spiders (Araneae: Salticidae). *Psyche* 72: 133–147.

Gertsch, W. J. 1934. Further notes on American spiders. *Am. Mus. Novit.* 726: 1–26.

Gertsch, W. J. 1939. A revision of the typical crab-spiders (Misumeninae) of America north of America. *Bull. Am. Mus. Nat. Hist.* 76: 277–442.

Gertsch, W. J. 1958. The spider family Diguetidae. *Am. Mus. Novit.* 1904: 1–24.

Gertsch, W. J. 1960. Descriptions of American spiders of the family Symphytognathidae. *Am. Mus. Novit.* 1981: 1–40.

Gertsch, W. J. 1974. The spider family Leptonetidae in North America. *J. Arachnol.* 1: 145–203.

Gertsch, W. J. 1979. *American Spiders.* Second Edition. Van Nostrand, Princeton, 274 pp.

Gertsch, W. J. 1982. The spider genera *Pholocophora* and *Anopsicus* (Araneae, Pholcidae) in North America, Central America and the West Indies. Assoc. Mexican Cave Stud. Bull. 8: 95–144/ Texas Mem. Mus. Bull. 28: 95–144.

Gertsch, W. J. 1984. The spider family Nesticidae (Araneae) in North America, Central America, and the West Indies. Texas Mem. Mus. Bull. 31: 1–91.

Gertsch, W. J. 1992. Distribution patterns and speciation in North American cave spiders with a list of the troglobites and revision of the *Cicurinas* of the subgenus *Circurella.* Texas Mem. Mus., Speleol. Monogr. 3: 75–122.

Gertsch, W. J., and F. Ennik. 1983. The spider genus *Loxosceles* in North America, Central America, and the West Indies (Araneae: Loxoscelidae). *Bull. Am. Mus. Nat. Hist.* 175: 264–360.

Gertsch, W. J., and S. Mulaik. 1936. Diagnoses of new southern spiders. *Am. Mus. Nov.* 851: 1– 21.

Gertsch, W. J., and S. Mulaik. 1940. The spiders of Texas I. *Bull. Am. Mus. Nat. Hist.* 77: 307–340.

Gertsch, W. J., and N. I. Platnick. 1980. A revision of the American spiders of the family Atypidae (Araneae, Mygalomorphae). *Am. Mus. Novit.* 2704: 1–39.

Gertsch, W. J., and M. Soleglad. 1966. The scorpions of the *Vejovis boreus* group (Subgenus *Paruroctonus*) in North America (Scorpionida, Vejovidae). *Am. Mus. Novit.* 2278: 1–54.

Gertsch, W. J., and M. Soleglad. 1972. Studies of North American scorpions of the genera *Uroctonus* and *Vejovis* (Scorpionida, Vejovidae). *Bull. Am. Mus. Nat. Hist.,* 148: 551–608.

Gertsch, W. J., and H. K. Wallace. 1935. Further notes on American Lycosidae. *Am. Mus. Novit.* 794: 1–22.

Greenstone, M. H. 1978. The numerical response to prey availability of *Pardosa ramulosa* (McCook) (Araneae: Lycosidae) and its relationship to the role of spiders in the balance of nature. *Symp. Zool. Soc. London.* 42: 183–193.

Greenstone, M. H. 1979a. Spider feeding behaviour optimizes dietary essential amino acid composition. *Nature* 282: 501–503.

Greenstone, M. H. 1979b. A line transect density index for wolf spiders (*Pardosa* spp.), and a note on the applicability of catch per unit effort methods to entomological study. *Ecol. Entomol.* 4: 23–29.

Greenstone, M. H. 1980. Contiguous allotopy of *Pardosa ramulosa* and *Pardosa tuoba* (Araneae: Lycosidae) in the San Francisco bay region, and its implications for patterns of resource partitioning in the genus. *Am. Midl. Nat.* 104: 305–311.

Griswold, C. E. 1987. A revision of the jumping spider genus *Habronattus* F.O.P.-Cambridge (Araneae; Salticidae), with phenetic and cladistic analyses. *Univ. California Publ. Entomol.* 107: 1–344.

Hair, J. A., and J. L. Bowman. 1986. Behavioral Ecology of *Amblyomma americanum* (L.). Chapter 18. In *Morphology, Physiology and Behavioral Biology of Ticks.* Ed. J. R. Sauer and J. A. Hair., Ellis Horwood Ltd., Chichester, UK.

Harwood, R. H. 1974. Predatory behavior of *Argiope aurantia* (Lucas). *Am. Midl. Nat.* 91: 130–139.

Hayes, J. L., and T. C. Lockley. 1990. Prey and nocturnal activity of wolf spiders (Araneae: Lycosidae) in cotton fields in the Delta region of Mississippi. *Environ. Entomol.* 19: 1512–1518.

Heiss, J . S., and R. T. Allen. 1986. The Gnaphosidae of Arkansas. Arkansas Agric. Exp. Stn. Bull. 885, Fayetteville.

Hoff, C. C. 1949. The Pseudoscorpions of Illinois. Bull. Ill. Nat. Hist. Survey. 24: 413–498.

Horner, N. V. 1972. *Metaphidippus galathea* as a possible biological control agent. *J. Kansas Entomol. Soc.* 45: 324–327.

Horner, N. V., and K. J. Starks. 1972. Bionomics of the jumping spider *Metaphidippus galathea. Ann. Entomol. Soc. Am.* 65: 602–607.

Jackson, R. R. 1982. The behavior of communicating in jumping spiders (Salticidae). In P. Witt and J. Rovner (eds). *Spider Communication: Mechanism and Ecological Significance:* 213–247. Princeton, N. J., Princeton, Univer. Press.

Jackson, R. R., and A. M. MacNab. 1989. Display behavior of *Corythalia canosa,* an ant-eating jumping spider (Araneae: Salticidae) from Florida, *N. Z. J. Zool.* 16: 169–183.

Janowski-Bell, M., and N. V. Horner. 1995. Of the movements of the Texas brown tarantula *Aphonopelma hentzi* (Araneae: Theraphosidae) using radio telemetry. Program and Abstracts of Papers. 98th Annual Meeting of the Texas Acad. of Science. Abstract 180.

Jones, S. E. 1936. The Araneida of Dallas County: preliminary note. *Field & Lab.* 4: 68–70.

Kagan, M. 1943. The Araneida found on cotton in central Texas. *Ann. Entomol. Soc. Am.* 36: 257–258.

Kaston, B. J. 1948. The spiders of Connecticut. St. Geol. and Nat. Hist. Surv. Bull. 70: 1–874.

Kaston, B. J. 1972. Web making by young *Peucetia. Notes Arachnol. Southwest* 3: 6–7.

Kaston, B. J. 1973. Four new species of *Metaphidippus,* with notes on related jumping spiders (Araneae: Salticidae) from the eastern and central United States. *Trans. Am. Micr. Soc.* 92: 106–122.

Kaston, B. J. 1978. *How to Know the Spiders.* 3rd ed. Wm. C. Brown Co., Dubuque, lowa, 272 pp.

Killebrew, D. W. 1982. *Mantispa* in a *Peucetia* egg case. *J. Arachnol.* 10: 281–282.

Killebrew, D. W., and N. B. Ford. 1985. Reproductive tactics and female body size in the green lynx spider *Peucetia viridans* (Araneae: Oxyopidae). *J. Arachnol.* 13: 375–382.

Knutson, A. E., and F. E. Gilstrap. 1989. Predators and parasites of the southwestern corn borer (Lepidoptera: Pyralidae) in Texas corn. *J. Kans. Entomol. Soc.* 62: 511–520.

Kuenzler, E. J. 1958. Niche relations of three species of Iycosid spiders. *Ecology* 39: 494–500.

Lee, R. C. P., M. Nyffeler, E. Krelina, and B. W. Pennycook. 1986. Acoustic communication in two spider species of the genus *Steatoda* (Araneae, Theridiidae). *Mitt. Schweiz. Entomol. Ges.* 59: 337–348.

Leech, R. 1972. A revision of the nearctic Amaurobiidae (Arachnida: Araneida). *Mem. Entomol. Soc. Can.* 84: 1–182.

Lehtinen, P. 1967. Classification of the cribellate spiders and some allied families. *Annales Zool. Fennici* 4: 199–468.

LeSar, C. D., and J. D. Unzicker. 1978. Life history, habits, and prey preferences of *Tetragnatha laboriosa* (Araneae: Tetragnathidae). *Environ. Entomol.* 7: 879–884.

Levi, H. W. 1954. Spiders of the genus *Euryopis* from North and Central America (Araneae,Theridiidae). *Am. Mus. Novit.* 1666: 1–48.

Levi, H. W. 1955a. The spider genera *Chrysso* and *Tidarren* in America (Araneae: Theridiidae). *J. New York Entomol. Soc.* 63: 59–81.

Levi, H. W. 1955b. The spider genera *Coressa* and *Achaearanea* in America north of Mexico (Araneae, Theridiidae). *Am. Mus. Novit.* 1718: 1–33.

Levi, H. W. 1956. The spider genera *Neottiura* and *Anelosimus* in America (Araneae: Theridiidae). *Trans. Am. Micr. Soc.* 75: 407–422.

Levi, H. W. 1957a. The spider genera *Enoplognatha, Theridion,* and *Paidisca* in America north of Mexico (Araneae, Theridiidae). *Bull. Am. Mus. Nat. Hist.* 112: 1–123.

Levi, H. W. 1957b.The spider genera *Crustulina* and *Steatoda* in North America, Central America, and the West Indies (Araneae: Theridiidae). *Bull. Mus. Comp. Zool.* 117: 367–424.

Levi, H. W. 1959a. The spider genus *Latrodectus* (Araneae, Theridiidae). *Trans. Am. Micr. Soc.* 78: 7–43.

Levi, H. W. 1959b. The spider genus *Coleosoma* (Araneae, Theridiidae). *Breviora* 110: 1–8.

Levi, H. W. 1968. The spider genera *Gea* and *Argiope* in America (Araneae: Araneidae). *Bull. Mus. Comp. Zool.* 136: 319–352.

Levi, H. W. 1970. The *ravilla* group of the orbweaver genus *Eriophora* in North America (Araneae: Araneidae). *Psyche* 77: 280–302.

Levi, H. W. 1971a. The *diadematus* group of the orb-weaver genus *Araneus* north of Mexico (Araneae: Araneidae). *Bull. Mus. Comp. Zool.* 141: 131–179.

Levi, H. W. 1971b. The orb-weaver genera *Singa* and *Hyposinga* in America (Araneae: Araneidae). *Psyche* 78: 229–256.

Levi, H. W. 1973. Small orb-weavers of the genus *Araneus* north of Mexico (Araneae: Araneidae). *Bull. Mus. Comp. Zool.* 145: 473–552.

Levi, H. W. 1974.The orb-weaver genera *Araniella* and *Nuctenea* (Araneae: Araneidae). *Bull. Mus. Comp. Zool.* 146: 291–316.

Levi, H. W. 1975. The American orb-weaver genera *Larinia, Cercidia* and *Mangora* north of Mexico (Araneae, Araneidae). *Bull. Mus. Comp. Zool.* 147: 101–135.

Levi, H. W. 1976. The orb-weaver genera *Verrucosa, Acanthepeira, Wagneriana, Acacesia, Wixia, Scoloderus,* and *Alpaida* north of Mexico (Araneae: Araneidae). *Bull. Mus. Comp. Zool.* 147: 351– 391.

Levi, H. W. 1977a. The American orb-weaver genera *Cyclosa, Metazygia,* and *Eustala* north of Mexico (Araneae, Araneidae). *Bull. Mus. Comp. Zool.* 148: 61–127.

Levi, H. W. 1977b. The orb-weaver genera *Metepeira, Kaira,* and *Aculepeira* in America north of Mexico (Araneae: Araneidae). *Bul. Mus. Comp. Zool.* 148: 185–238.

Levi, H. W. 1978. The American orb-weaver genera *Colphepeira, Micrathena,* and *Gasteracantha* north of Mexico (Araneae, Araneidae). *Bull. Mus. Comp. Zool.* 148: 417–442.

Levi, H. W. 1980. The orb-weaver genus *Mecynogea,* the subfamily Metinae, and the genera *Pachygnatha, Glenognatha,* and *Azilia* of the subfamily Tetragnathinae north of Mexico (Araneae: Araneidae). *Bull. Mus. Comp. Zool.* 149: 1–74.

Levi, H. W. 1981. The American orb-weaver genera *Dolichognatha* and *Tetragnatha* north of Mexico (Araneae: Araneidae, Tetragnathinae). *Bull. Mus. Comp. Zool.* 149: 271 –318.

Levi, H. W., and D. E. Randolph. 1975. A key and checklist of American spiders of the family Theridiidae north of Mexico (Araneae). *J. Arachnol.* 3: 31–51.

Levi, H. W., L. R. Levi, and H. S. Zim. 1990. *Spiders and Their Kin.* Golden Press, New York, 160 pp.

Liao, H-t., M. K. Harris, F. E. Gilstrap, D. A. Dean, C. W. Agnew, G. J. Michels, and F. Mansour. 1984. Natural enemies and other factors affecting seasonal abundance of the blackmargined aphid on pecan. *Southwest. Entomol.* 9: 404–420.

MacKay, W. P. 1982. The effect of predation of western widow spiders (Araneae: Theridiidae) on harvester ants (Hymenoptera: Formicidae). *Oecologia* 53: 406–411.

MacKay, W. P., and S. B. Vinson. 1989. Evaluation of the spider *Steatoda triangulosa* (Araneae: Theridiidae) as a predator of the red imported fire ant (Hymenoptera: Formicidae). *J. N. Y. Entomol. Soc.* 97: 232–233.

Maddison, W. P. 1996. *Pelegrina franganillo* and other jumping spiders formerly placed in the genus *Metaphidippus* (Araneae: Salticidae). *Bull. Mus. Comp. Zool.* 154 (4): 215–360.

McCaffrey, J. P., and R. L. Horsburgh. 1980. The spider fauna of apple trees in central Virginia. *Environ. Entomol.* 9: 247–252.

McDaniel, B. 1979. *How to Know the Mites and Ticks.* Wm. C. Brown Company, Dubuque, Iowa, 335 pp.

McIlveen, G., Jr., P. D. Teel, and P. J. Hamman. 1990. Preventing Lyme disease. Texas Agricul. Extension Serv. B–1660. 8 pp.

Muma, M. H. 1951. The arachnid order Solpugida in the United States. *Bull. Am. Mus. Nat. Hist.,* 97: 35–141.

Muma, M. H. 1953. A study of the spider family Selenopidae in North America, Central America, and the West Indies. *Am. Mus. Novit.* 1619: 1–55.

Muma, M. H. 1962. The arachnid order Solpugida in the United States, Supplement I. *Am. Mus. Novit.* 2092: 1–44.

Muma, M. H. 1975. Spiders in Florida citrus groves. *Florida Entomol.* 58: 83–90.

Muma, M. H., and W. J. Gertsch. 1964. The spider family Uloboridae in North America north of Mexico. *Am. Mus. Novit.* 2196: 1–43.

Muniappan, K., and H. L. Chada. 1970. Biology of the crab spider, *Misumenops celer. Ann. Entomol. Soc. Am.* 63: 1718–1722.

Nyffeler, M., and G. Benz. 1981. Some observations on the feeding ecology of the wolf spider *Pardosa lugubris* (Walck.). *Deutsche Entomol. Z.,* Berlin 28: 297–300 (in German).

Nyffeler, M., and G. Benz. 1988. Feeding ecology and predatory importance of wolf spiders (*Pardosa* spp.) (Araneae, Lycosidae) in winter wheat fields. *J. Appl. Entomol.* 106: 123–134.

Nyffeler, M., and R. G. Breene. 1990. Evidence of low daily food consumption by wolf spiders in meadowland and comparison with other cursorial hunters. *J. Appl. Entomol.* 110: 73–81.

Nyffeler, M., D. A. Dean, and W. L. Sterling. 1986a. Feeding habits of the spiders *Cyclosa turbinata* (Walckenaer) and *Lycosa rabida* Walckenaer. *Southwest. Entomol.* 11: 195–201.

Nyffeler, M., C. D. Dondale, and J. H. Redner. 1986b. Evidence for displacement of a North American spider, *Steatoda borealis* (Hentz) by the European species *S. bipunctata* (Linnaeus) (Araneae: Theridiidae). *Can. J. Zool.* 64: 867–874.

Nyffeler, M., D. A. Dean, and W. L. Sterling. 1987a. Evaluation of the importance of the striped lynx spider, *Oxyopes salticus* (Araneae: Oxyopidae), as a predator in Texas cotton. *Environ. Entomol.* 14: 1114–1123.

Nyffeler, M., D. A. Dean, and W. L. Sterling. 1987b. Feeding ecology of the orb-weaving spider *Argiope aurantia* (Araneae: Araneidae), in a cotton agroecosystem. *Entomophaga* 32: 367–375.

Nyffeler, M., D. A. Dean, and W. L. Sterling. 1987c. Predation by green lynx spider, *Peucetia viridans* (Araneae: Oxyopidae), inhabiting cotton and woolly croton plants in east Texas. *Environ. Entomol.* 16: 355–359.

Nyffeler, M., D. A. Dean, and W. L. Sterling. 1988. Prey records of the web-building spiders *Dictyna segregata* (Dictynidae), *Theridion australe* (Theridiidae), *Tidarren haemorrhoidale* (Theridiidae), and *Frontinella pyramitela* (Linyphiidae) in a cotton agroecosystem. *Southwest. Naturalist* 33: 215–218.

Nyffeler, M., D. A. Dean, and W. L. Sterling. 1988. The southern black widow spider, *Latrodectus mactans* (Araneae, Theridiidae), as a predator of the red imported fire ant, *Solenopsis invicta* (Hymenoptera, Formicidae), in Texas cotton fields. *J. Appl. Entomol.* 106: 52–57.

Nyffeler, M., D. A. Dean, and W. L. Sterling. 1989. Prey selection and predatory importance of orb-weaving spiders (Araneae: Araneidae, Uloboridae) in Texas cotton. *Environ. Entomol.* 18: 373–380.

Nyffeler, M., R. G. Breene, D. A. Dean, and W. L. Sterling. 1990. Spiders as predators of arthropod eggs. *J. Appl. Entomol.* 109: 490–501.

Nyffeler, M., D. A. Dean, and W. L. Sterling. 1992a. Spiders associated with lemon horsemint (*Monarda citriodora* Cervantes) in east central Texas. Texas Agric. Exp. Stn. Bull. 1707, College Station.

Nyffeler, M., W. L. Sterling, and D. A. Dean. 1992b. Impact of the striped lynx spider (Araneae: Oxyopidae) and other natural enemies on the cotton fleahopper (Hemiptera: Miridae) in Texas cotton. *Environ. Entomol.* 21: 1178–1188.

Nyffeler, M., D. A. Dean, and W. L. Sterling. 1992c. Diets, feeding specialization, and predatory role of two lynx spiders, *Oxyopes salticus* and *Peucetia viridans* (Araneae: Oxyopidae), in a Texas cotton agroecosystem. *Environ. Entomol.* 21: 1457–1465.

Nyffeler, M., W. L. Sterling, and D. A. Dean. 1994. How spiders make a living. *Environ. Entomol.* 23: 1357–1367.

Opell, B. D. 1979. Revision of the genera and tropical American species of the spider family Uloboridae. *Bull. Mus. Comp. Zool.* 148: 443–549.

Opell, B. D., and J. A. Beatty. 1976. The nearctic Hahniidae (Arachnida: Araneae). *Bull. Mus. Comp. Zool.* 147: 393–433.

Pase, H. A., and D. T. Jennings. 1978. Bite by the spider *Trachelas volutus* Certsch (Araneae: Clubionidae). *Toxicon* 16: 96–98.

Peck, W. B. 1981. The Ctenidae of temperate zone North America. *Bull. Am. Mus. Nat. Hist.* 170: 157–169.

Peck, W. B., and W. H. Whitcomb. 1970. Studies on the biology of a spider *Chiracanthium inclusum* (Hentz). Arkansas Agric. Exp. Stn. Bull. 753, Fayetteville.

Peckham, G. W., and E. G. Peckham. 1909. Revision of the Attidae of North America. Trans. Wisconsin Acad. Sci. Arts and Letters 6: 355–646.

Platnick, N. I. 1974. The spider family Anyphaenidae in America north of Mexico. *Bull. Mus. Comp. Zool.* 146: 205–266.

Platnick, N. I. 1986. On the tibial and patellar glands, relationships, and American genera of the spider family Leptonetidae (Arachnida, Araneae). *Am. Mus. Novit.* 2855: 1–16.

Platnick, N. I. 1989. Advances in spider taxonomy 1981–1987: a supplement to Brignoli's *A Catalogue of the Araneae Described Between 1940 and 1981.* Manchester Univ. Press, Manchester, England, 673 pp.

Platnick, N. I. 1993. Advances in Spider Taxonomy 1982–1991 with Synonymies and Transfers 1940–1980. Manchester Univ. Press, Manchester, England, 846 pp.

Platnick, N. I., and M. U. Shadab. 1974a. A revision of the *tranguillus* and *speciosus* groups of the spider genus *Trachelas* (Araneae, Clubionidae) in North and Central America. *Am. Mus. Novit.* 2553: 1–34.

Platnick, N. I., and M. U. Shadab. 1974b. A revision of the *bispinosus* and *bicolor* groups of the spider genus *Trachelas* (Araneae, Clubionidae) in North and Central America and the West Indies. *Am. Mus. Novit.* 2560: 1–34.

Platnick, N. I., and M. U. Shadab. 1975. A revision of the spider genus *Gnaphosa* (Araneae, Gnaphosidae) in America. *Bull. Am. Mus. Nat. Hist.* 155: 1–66.

Platnick, N. I., and M. U. Shadab. 1976. A revision of the spider genera *Rachodrassus, Sosticus,* and *Scopodes* (Araneae, Gnaphosidae) in North America. *Am. Mus. Novit.* 2594: 1–33.

Platnick, N. I., and M. U. Shadab. 1980. A revision of the North American spider genera *Nodocion, Litopyllus,* and *Synaphosus* (Araneae, Gnaphosidae). *Am. Mus. Novit.* 2691: 1–26.

Platnick, N. l., and M. U. Shadab. 1981. A revision of the spider genus *Sergiolus* (Araneae, Gnaphosidae). *Am. Mus. Novit.* 2717: 1–41.

Platnick, N. I., and M. U. Shadab. 1982. A revision of the American spiders of the genus *Drassyllus* (Araneae, Gnaphosidae). *Bull. Am. Mus. Nat. Hist.* 173: 1–97.

Platnick, N. I., and M. U. Shadab. 1988. A revision of the American spiders of the genus *Micaria* (Araneae, Gnaphosidae). *Am. Mus. Novit.* 2916: 1–64.

Polis, G. A. 1990. *The Biology of Scorpions.* Stanford University Press. Stanford, California, 587 pp.

Randall, J. B. 1977 (1978). New observations of maternal care exhibited by the green lynx spider, *Peucetia viridans* Hentz (Araneida: Oxyopidae). *Psyche* 84: 286–291.

Randall, J. B. 1978. The use of femoral spination as a key to instar determination in the green lynx spider, *Peucetia viridans* (Hentz) (Araneida: Oxyopidae). *J. Arachnol.* 6: 147– 153.

Randall, J. B. 1982. Prey records of the green lynx spider, *Peucetia viridans* (Hentz) (Araneae, Oxyopidae). *J. Arachnol.* 10: 19–22.

Rapp, W. F. 1984. The spiders of Galveston Island (Texas). *Novit. Arthropodae* 2: 1–10.

Reddell, J. R. 1965. A checklist of the cave fauna of Texas. I. The Invertebrata (exclusive of Insecta). *Texas J. Sci.* 17: 143–187.

Reddell, J. R. 1970. A checklist of the cave fauna of Texas. IV. Additional records of Invertebrata (exclusive of Insecta). *Texas J. Sci.* 21: 389–415.

Reiskind, J. 1969. The spider subfamily Castianeirinae of North and Central America (Araneae, Clubionidae). *Bull. Mus. Comp. Zool.* 138: 163–325.

Richman, D. B. 1978. Key to the jumping spiders (Salticid) genera of North America. *Peckhamia* 1: 77–81.

Richman, D. B. 1989. A revision of the genus *Hentzia* (Araneae, Salticidae). *J. Arachnol.* 17: 285–344.

Richman, D. B., and B. Cutler. 1978. A list of the jumping spiders (Araneae: Salticidae) of the United States and Canada. *Peckhamia* 1: 82–110.

Richman, D. B., and R. R. Jackson. 1992. A review of the ethology of jumping spiders (Araneae, Salticidae). Bull. Br. Arachnol. Soc. 9: 33–37.

Richman, D. B., and W. H. Whitcomb. 1981. The ontogeny of *Lyssomanes viridis* (Walckenaer) (Araeneae: Salticidae) on *Magnolia grandifolia* L. *Psyche* 88: 127–133.

Roach, S. H. 1987. Observations on feeding and prey selection by *Phidippus audax* (Araneae: Salticidae). *Environ. Entomol.* 16: 1098–1102.

Robinson, M. H. 1969. Predatory behaviour of *Argiope argentata* (Fabricius). *Am. Zool.* 9: 161–174.

Rogers, C. E., and N. V. Horner. 1977. Spiders of guar in Texas and Oklahoma. *Environ. Entomol.* 6: 523–524.

Roth, V. D. 1993. *Spider Genera of North America.* Am. Arachnol. Soc. 203 pp.

Rowland, J. M., and J. A. L. Cooke. 1973. Systematics of the arachnid order Uropygida (=Thelyphonida). *J. Arachnol.* 1: 55–71.

Rowland, J. M., and J. R. Reddell. 1976. Annotated checklist of the arachnid fauna of Texas (excluding Acarida and Araneida). Occasional Papers, the Museum Texas Tech Univ. No. 38, 25 pp.

Sabath, L. E. 1969. Color change and life history observations of the spider *Gea heptagon* (Araneae: Araneidae). *Psyche* 76: 367–374.

Sauer, R. J., and N. I. Platnick. 1972. The crab spider genus *Ebo* (Araneida: Thomisidae) in the United States and Canada. *Can. Entomol.* 104: 35–60.

Shear, W. A. 1970. The spider family Oecobiidae in North America, Mexico, and the West Indies. *Bull. Mus. Comp. Zool.* 140: 129–164.

Shelley, R. M., and W. D. Sissom. 1995. Distributions of the scorpions *Centruroides vittatus* (Say) and *Centruroides hentzi* (Banks) in the United States and Mexico (Scorpiones, Buthidae). *J. Arachnol.* 23: 100–110.

Sissom, W. D. 1990. Systematics of *Vaejovis dugesi* Pocock, with descriptions of two new related species (Scorpiones, Vaejovidae). *Southwest. Nat.* 35: 47–53.

Sissom, W. D. , and O. F. Francke. 1981. Scorpions of the genus *Paruroctonus* from New Mexico and Texas (Scorpiones, Vaejovidae). *J. Arachnol.* 9: 93–108.

Sissom, W. D. , and O. F. Francke. 1985. Redescriptions of some poorly known species of the *Nitidulus* group of the genus *Vaejovis* (Scorpiones, Vaejovidae). *J. Arachnol.* 13: 243–266.

Smith, A. M. 1994. *A Study of the Theraphosidae Family from North America.* Fitzgerald Publishing, P.O. Box 804, London SE 13 7JF England, 196 pp.

Sterling, W. L. 1982. Predacious insects and spiders. In Identification, biology and sampling of cotton insects. Bohmfalk, G. T., R. E. Frisbie, W. L. Sterling, R. B. Metzer, and A. E. Knutson. Texas Agric. Ext. Serv. B–933, 44 pp.

Stockwell, S. A. 1986. The Scorpions of Texas (Arachnida, Scorpiones). Unpublished Master's Thesis. Texas Tech University.

Teel, P. D. 1985. Ticks. In Williams, R. E., R. D. Hall, A. B. Broce, and P. J. Scholl. *Livestock Entomology,* John Wiley & Sons, New York.

Turner, M. 1979. Diet and feeding phenology of the green lynx spider, *Peucetia viridans* (Araneae: Oxyopidae). *J. Arachnol.* 7: 149–154.

Uetz, G. W. 1973. Envenomation by the spider *Trachelas tranquillus* (Hentz) (Araneae: Clubionidae). *J. Med. Entomol.* 10: 227.

Uetz, G. W., and J. M. Biere. 1980. Prey of *Micrathena gracilis* (Walckenaer) (Araneae: Araneidae) in comparison with artificial webs and other trapping devices. Bull. Brit. Arachnol. Soc. 5: 101–107.

Valerio, C. E. 1981. Spitting spiders (Araneae, Scytodidae, *Scytodes*) from Central America. *Bull. Am. Mus. Nat. Hist.* 170: 80–89.

Vogel, B. R. 1970a. Bibliography of Texas spiders. *Armadillo Papers* 2: 1–36.

Vogel, B. R. 1970b. Taxonomy and morphology of the *sternalis* and *falcifera* species groups of *Pardosa* (Araneida: Lycosidae). *Armadillo Papers* 3: 1–31.

Vogel, B., and C. Durden. 1972. New year greetings. Some of the spiders around our house. Privately Printed. 6 pp.

Wallace, H. K., and H. Exline. 1978. Spiders of the genus *Pirata* in North America, Central America and the West Indies (Araneae: Lycosidae). *J. Arachnol.* 5: 1–112.

Weems, Jr., H. V., and W. H. Whitcomb. 1977. The green lynx spider, *Peucetia viridans* (Hentz) (Araneae: Oxyopidae). Florida Dept. Agric. Cons. Serv. DPI Entomol. Circ. 181: 1–4.

Weygoldt, P. 1969. *The Biology of Pseudoscorpions.* Harvard Univ. Press. Cambridge, Massachusetts.

Wheeler, A. G., Jr. 1973. Studies on the arthropod fauna of alfalfa. V. Spiders (Araneida). *Can. Entomol.* 105: 425–432.

Whitcomb, W. H. 1962. Egg-sac construction and oviposition of the green lynx spider *Peucetia viridans* (Oxyopidae). *Southwest. Nat.* 7: 198–201.

Whitcomb, W. H., and K. Bell. 1964. Predacious insects, spiders, and mites of Arkansas cotton fields. Arkansas Agric. Exp. Stn. Bull. 690, Fayetteville.

Whitcomb, W. H., and R. Eason. 1965. The mating behavior of *Peucetia viridans* (Araneida: Oxyopidae). *Florida Entomol.* 48: 163–167.

Whitcomb, W. H., and R. Eason. 1967. Life history and predatory importance of the striped lynx spider (Araneida: Oxyopidae). Arkansas Acad. Sci. Proc. 21: 54–58.

Whitcomb, W. H., H. Exline, and R. C. Hunter. 1963. Spiders of the Arkansas cotton field. *Ann. Entomol. Soc. Am.* 56: 653–660.

Whitcomb, W. H., M. Hite, and R. Eason. 1966. Life history of the green lynx spider, *Peucetia viridans* (Araneida: Oxyopidae). *J. Kansas Entomol. Soc.* 39: 259–267.

Willey, M. B., and P. H. Adler. 1989. Biology of *Peucetia* (Araneae, Oxyopidae) in South Carolina, with special reference to predation and maternal care. *J. Arachnol.* 17: 275–284.

Williams, S. C. 1968. Scorpions from Northern Mexico: Five new species of *Vejovis* from Coahuila, Mexico. Occasional Papers California Acad. Sci., 68: 1–24.

Williams, H. E., R. G. Breene, and R. S. Rees. 1986. The black widow spider. University of Tennessee Agric. Ext. PB1193, 12 pp.

Woods, M. W., and R. C. Harrell. 1976. Spider populations of a southeast Texas rice field. *Southwest. Nat.* 21: 37–48.

Yeargan, K. V. 1975. Prey and periodicity of *Pardosa ramulosa* (McCook) in alfalfa. *Environ. Entomol.* 4: 137–141.

Young, O. P. 1989a. Field observations of predation by *Phidippus audax* (Araneae: Salticidae) on arthropods associated with cotton. *J. Entomol. Sci.* 24: 266–273.

Young, O. P. 1989b. Interactions between the predators *Phidippus audax* (Araneae: Salticidae) and *Hippodamia convergens* (Coleoptera: Coccinellidae) in cotton and the laboratory. *Entomol. News* 100: 43–47.

Young, O. P., and G. B. Edwards. 1990. Spiders in United States field crops and their potential effect on crop pests. *J. Arachnol.* 18: 1–27.

Young, O. P., and T. C. Lockley. 1985. The striped lynx spider, *Oxyopes salticus* (Araneae: Oxyopidae), in agroecosystems. *Entomophaga* 30: 329–346.

Young, O. P., and T. C. Lockley. 1986. Predation of striped lynx spider, *Oxyopes salticus* (Araneae: Oxyopidae), on tarnished plant bug, *Lygus lineolaris* (Heteroptera: Miridae): a laboratory evaluation. *Ann. Entomol. Soc. Am.* 79: 879–883.

Zolnerowich, G., and N. V. Horner. 1985. Gnaphosid spiders of north-central Texas (Araneae, Gnaphosidae). *J. Arachnol.* 13: 79–85.

Index

The letter P preceding numbers indicates color plate pages.

About the Author

John A. Jackman, Ph.D., is a professor and extension entomology specialist at the Texas A&M University System, College Station, Texas. He is a member of the Entomological Society of America.